KA 0427616 7

CONTESTING RURALITY

Perspectives on Rural Policy and Planning

Series Editors:
Andrew Gilg
University of Exeter, UK
Keith Hoggart
King's College London, UK
Henry Buller
Cheltenham College of Higher Education, UK
Owen Furuseth
University of North Carolina, USA
Mark Lapping
University of South Maine, USA

Other titles in the series

**Equity, Diversity and Interdependence
Reconnecting Governance and People through Authentic Dialogue**
ISBN 0 7546 3521 X

Women in the European Countryside
Edited by Henry Buller and Keith Hoggart
ISBN 0 7546 3946 0

**Mapping the Rural Problem in the Baltic Countryside
Transition Processes in the Rural Areas of Estonia, Latvia and Lithuania**
Edited by Ilkka Alanen
ISBN 0 7546 3434 5

Geographies of Rural Cultures and Societies
Edited by Lewis Holloway and Moya Kneafsey
ISBN 0 7546 3571 6

Big Places, Big Plans
Edited by Mark B. Lapping and Owen J. Furuseth
ISBN 0 7546 3586 4

Contesting Rurality
Politics in the British Countryside

MICHAEL WOODS
University of Wales, Aberystwyth

Routledge
Taylor & Francis Group
LONDON AND NEW YORK

First published 2005 by Ashgate Publishing

Published 2016 by Routledge
2 Park Square, Milton Park, Abingdon, Oxon OX14 4RN
711 Third Avenue, New York, NY 10017, USA

Routledge is an imprint of the Taylor & Francis Group, an informa business

Copyright © 2005 Michael Woods

Michael Woods has asserted his right under the Copyright, Designs and Patents Act, 1988, to be identified as author of this work.

All rights reserved. No part of this book may be reprinted or reproduced or utilised in any form or by any electronic, mechanical, or other means, now known or hereafter invented, including photocopying and recording, or in any information storage or retrieval system, without permission in writing from the publishers.

Notice:
Product or corporate names may be trademarks or registered trademarks, and are used only for identification and explanation without intent to infringe.

British Library Cataloguing in Publication Data
Woods, Michael
    Contesting rurality : politics in the British countryside.
    - (Perspectives on rural policy and planning)
    1.Sociology, Rural - Great Britain 2.Rural development - Great Britain 3.Local government - Great Britain 4.Great Britain - Rural conditions 5.Great Britain - Politics and government - 1997-
    I.Title
    307.7'2'0941

Library of Congress Cataloging-in-Publication Data
Woods, Michael.
    Contesting rurality : politics in the British countryside / by Michael Woods.-- 1st ed.
        p. cm. -- (Perspectives on rural policy and planning)
    Includes bibliographical references and index.
    ISBN 0-7546-3025-0
    1. Sociology, Rural--Great Britain. 2. Great Britain--Rural conditions. I. Title. II. Series.

HT443.G7W66 2005
307.72'0941--dc22

2004027705

ISBN 9780754630258 (hbk)
ISBN 9781138277519 (pbk)

Transferred to Digital Printing 2014

Contents

List of Figures		*vi*
List of Tables		*vii*
Acknowledgements		*viii*
List of Abbreviations		*ix*
1	The Strange Awakening of Rural Britain	1
2	The Changing Balance of Local Power in the Countryside	23
3	Contemporary Rural Elites	47
4	National Politics and Rural Representation	84
5	The Countryside Alliance and Rural Protest	101
6	Agricultural Politics	131
7	Developing the Countryside: Discourses and Dissent	162
Bibliography		*193*
Index		*203*

List of Figures

Figure 3.1	The traditional county establishment network	55
Figure 3.2	The agricultural elite network	56
Figure 3.3	The emerging 'Liberal' elite network	58
Figure 3.4	Networks through public schools and London clubs	66
Figure 3.5	The Lieutenancy Dining Club network	68
Figure 3.6	Connections between High Sheriffs	71
Figure 3.7	The Taunton business elite network	76
Figure 5.1	Flags on display in the Liberty and Livelihood March	118
Figure 5.2	Placard at the Liberty and Livelihood March	126
Figure 5.3	Green Union Jack logo of the Real Countryside Alliance, attached to a rural road sign	128
Figure 6.1	Farmers' protests in Britain, December 1997 – March 2000	147
Figure 6.2	Protest against disposal of infected carcasses at Sennybridge, Powys	158

List of Tables

Table 1.1	Employment in non-metropolitan districts of England and Wales, 1951 and 1991	16
Table 1.2	Population change in rural and urban districts of England	16
Table 3.1	Composition of Taunton Deane Borough Council, 1973-1995	52
Table 3.2	Composition of West Somerset District Council, 1973-1995	52
Table 3.3	Representation of key groups on local councils, 1996	53
Table 4.1	Party support in rural constituencies, 1992 and 1997	97
Table 4.2	Change in Labour vote 1997-2001 by constituency type	100
Table 5.1	Chartered coaches to the Liberty and Livelihood March by organizer	105
Table 5.2	Liberty and Livelihood March, chartered coaches, passengers on chartered trains and beacons by county	106
Table 5.3	Involvement of Countryside Alliance members in campaigning and social activities	110
Table 6.1	Change to average farmgate price paid to farmers in Britain	143
Table 6.2	Average share of retail price received by farmers	143
Table 6.3	Premises infected with Foot and Mouth, by county	156
Table 7.1	Public attitudes to development in the British countryside	163
Table 7.2	Proposed housing development in Buckinghamshire, 1991-2011	179

Acknowledgements

This book is based on a series of research activities conducted over the last ten years. The first part of the book, exploring local politics in Somerset, reports on doctoral research undertaken at the University of Bristol between 1993 and 1996, funded by an Economic and Social Research Council Research Studentship. I would like to acknowledge the support and company of staff and postgraduate students at Bristol during this period, and particularly the guidance and advice of my supervisor, Paul Cloke. The second part of the book, focussing on rural issues in national politics, draws on research undertaken since 1996 at the University of Wales, Aberystwyth. Again, I am grateful to my colleagues for the convivial and supportive intellectual atmosphere and particularly for collaboration in research and writing. Some of the research presented here has been funded by the University of Wales Aberystwyth Research Fund.

All photographs are from the author's private collection.

List of Abbreviations

BASC	British Association for Shooting and Conservation
BFSS	British Field Sports Society
BSE	Bovine Spongiform Encephalopathy
CAP	Common Agricultural Policy
CJD	Creutzfeldt Jakob Disease
CLA	Country Landowners' Association (now Country Land and Business Association)
CND	Campaign for Nuclear Disarmament
COPA	Comité des Organizations Professionelles Agricoles
CPRE	Council for the Protection of Rural England (now Campaign to Protect Rural England)
DEFRA	Department for the Environment, Food and Rural Affairs
DETR	Department for the Environment, Transport and the Regions
EC	European Community
EEC	European Economic Community
EU	European Union
FFA	Farmers for Action
FMD	Foot and Mouth Disease
FUW	Farmers' Union of Wales
MAFF	Ministry of Agriculture, Fisheries and Food
MP	Member of Parliament
NFU	National Farmers' Union
NHS	National Health Service
RSPCA	Royal Society for the Protection of Cruelty to Animals
WTO	World Trade Organization

Abbreviations used in the network diagrams (not listed above):

CC	County Council(lor)
Ch	Chairperson
CHC	Member, Community Health Council
Co MD	Company Managing Director
DC	District Council(lor)
DG	Director General
DL	Deputy Lieutenant
ENPA	Member, Exmoor National Park Authority
f	Former (prefix)
Gov	School Governor
HS	High Sheriff
JP	Magistrate (Justice of the Peace)

MEP	Member of the European Parliament
NHS Tr	Member, NHS Trust
NRA	Member, National Rivers Authority Committee
PC	Parish Council(lor)
RDC	Rural District Councillor
SCC	Somerset County Council
SHC	Somerset Health Commission
TDBC	Taunton Deane Borough Council
TEC	Member, Training and Enterprise Council
Tr	Treasurer
VC	Vice Chair
VLL	Vice Lord Lieutenant
VP	Vice President

Chapter 1
The Strange Awakening of Rural Britain

Introduction

In April 1991 a ripple of excitement ran through the small Somerset village of Hinton St George as political machinations and alleged dirty tricks in the elections to the parish council briefly brought notoriety for the community in the national press. According to a report in *The Independent* newspaper the controversy had been sparked by the decision of all seven serving members of the council to stand down at the election. Action to fill the vacuum had been taken by one senior villager, an elderly widow 'whose husband used to run the Suez canal' (Dunn, 1991, p. 3.), who took it upon herself to recruit seven new candidates – a strategy that is not unusual in the politics of English parish councils (see Edwards and Woods, 2004). In Hinton, however, the apparently secretive way in which the recruitment was undertaken led to accusations of 'plotting' from the retiring councillors.

As the *Independent* journalist, Peter Dunn, observed, the story said 'a lot about the changes in English rural life' (Dunn, 1991, p. 3). Hinton St George is, as Dunn described, a former estate village where 'older residents still remember touching their caps to the local squire, Lord Poulett' (*ibid.*). Yet, the last Earl had died twenty years previously and the 'big house' sold by the family. As the 'old village' had 'drifted away', excluded by spiraling property prices, middle class in-migrants had moved in. Alongside the neighbourhood watch scheme and the cheese and wine parties in aid of the lifeboat charity, advertised in the window of the 'Personal Service Stores', the newcomers had also introduced their own ideas about the management of the village:

> Incomers, retired or simply seeking a better life, have transformed their adopted villages from communities of poorly-paid estate workers to places of chocolate box tranquility. They brought in street lighting with signs on the lamp posts warning old ladies about their incontinent dogs. They demanded, unsuccessfully, that the parish council should change the name of Gas Lane to something less offensive. A feminist element complained the village snooker hall, known as The Men's Reading Room, was a sexist affront. (*Ibid.*).

Yet, the Hinton story is not a tale of conflict between locals and incomers. All of the central characters in the dispute were incomers, testifying to the complexity of the reconstituted British countryside in which simplistic dichotomies belie the heterogeneity of in-migrants. Rather, the tensions in Hinton involved a challenge to the leadership of established upper middle class incomers by more recent, service class, in-migrants with 'a sense of mission', who accused the older elite of 'only [paying] lip service to issues like low-cost housing for needier families' (*ibid.*). In this way the newspaper article about Hinton St George could have been written, with a few changes to the details, about any number of rural communities in Britain that experienced significant social and economic restructuring in the late twentieth century. The key themes in the Hinton story are leitmotifs across the British countryside – the decline of paternalism and the agrarian economy, the effects of counterurbanization, class recomposition and gentrification, the influence of particular individuals who orchestrate rural community politics, the contrasting ideas of rural life mobilized by different factions, and even the symbolic importance of village institutions – as the dispute in Hinton revived old feuds between 'committees running the village hall and primary school' (*ibid.*).

For all its rich comment on rural community change, though, perhaps the most significant statement in the *Independent* article is the opening paragraph:

> Hinton St George, a serene Somerset village south-west of Yeovil, seems an unlikely setting for infighting you would expect in Tower Hamlets or some other troubled corner of metropolitan Britain. (Dunn, 1991, p. 3).

Thus even whilst describing a rural political conflict the myth is perpetuated that the countryside is a stable, ordered, virtually apolitical society. Dunn was not alone in reproducing this perception. Whilst a robust political scene has long been accepted as part of rural life in countries such as France and, to a lesser extent, the United States, in Britain 'politics' have been identified in the popular imagination with urban society. Even the routine practice of governing rural territories has been somehow seen as devoid of politics. Madgwick (1974), for example, in his study of Cardiganshire in *The Politics of Rural Wales*, quotes one party secretary who told his research team, 'if you give me a tenth of your grant I'll try to see that there's some politics going on for you to study' (p. 11).

Fast forward now to September 2003. The streets of central London are filled with more than 400,000 protesters participating in the Liberty and Livelihood March. According to the event's organizers, the Countryside Alliance – a organization that with 90,000 members had grown in the space of five years to become one of Britain's most prominent pressure groups – the demonstrators had come to the capital from all over rural Britain. There were 39 coaches from Cumbria, 11 from Pembrokeshire, 36 from Cornwall. From Lincolnshire came 44 coaches and a chartered train, from Durham, 15 coaches and three trains. As with the preceding Countryside Rally in 1997 and Countryside March in 1998, the overriding motivation for the marchers was the threatened ban on the hunting of wild mammals with hounds, yet mingled in the crowd were individuals who voiced other rural concerns – about the state of agriculture, about the closure of rural services, about rural housing and about the introduction of unfettered public access to open countryside.

In contrast to the earlier demonstrations, however, there bubbled beneath the surface of the Liberty and Livelihood March a mood of defiance and belligerence that had already found expression in the activities of militant pro-hunting groups such as the Countryside Action Network – which had blockaded motorways – and the 'Real Countryside Alliance' – which had placed a giant *papier mache* huntsman on the Uffingham White Horse and hung a pro-hunting banner from the wings of the Angel of the North. Moreover, these gestures were mirrored by the direct action tactics of militant farmers, notably the Farmers for Action group, who had played a leading role in the blockade of fuel refineries and depots in September 2000 and who had subsequently targeted dairies and food processing plants. At the same time, elsewhere in the British countryside, local protests were being mobilized against proposed housing developments and new roads, and to protect village schools and post offices. Some of the communities affected were still recovering from the consequences of the Foot and Mouth epidemic in 2001, an agricultural crisis that had attracted huge amounts of political and media attention and had generated numerous local conflicts over the handling of the outbreak and the designation of disposal sites for culled livestock.

Against this visible background of apparent rural discontent a new arrival to Britain browsing through domestic news reports from any of the last five years would find the idea that the British countryside is an 'apolitical space' frankly incomprehensible. In the course of just a decade the concerns of rural Britain have moved to the centre of political discourse and activity. This book seeks to document, analyze and explain this 'strange political awakening' of rural Britain. It does so by knitting together the myriad different processes of change that have contributed to this transition, operating at different scales and in different contexts. Thus the early chapters examine the impact of social and economic restructuring on the local political structures of rural areas. The book proceeds to trace the 'scaling-up' of rural conflicts from the local to the national political arena, investigating the growing prominence of rural issues in national politics and the policy developments that have both responded to and contributed to this process. Finally, the later chapters explore in detail some of the key areas of conflict in contemporary rural politics, including agriculture, hunting, housing development, windfarms and road-building.

However, before the impression is created that this book is about the politicization of the British countryside, it is perhaps necessary to first return to the notion of the 'apolitical countryside' and to critically examine its providence and its effect in disguising the historically entrenched exercise of power in rural society.

The Myth of the Apolitical Countryside

The British countryside is a land of myths. Despite the periodic attempt of academics and government officials to establish and impose 'objective' definitions of rural space, the 'countryside' as an idea has always been at its most vivid and most powerful in its cultural construction in the popular imagination. The 'rural idyll' myth, for example, has been a central tenet of British culture, influencing leisure and residential patterns. The attraction of the 'rural idyll' has been a strong driver in the process of counterurbanization and has informed attitudes towards countryside conservation. The 'rural idyll' also contributed to the development of separate myths that identified the countryside with national identity in England, Scotland and Wales. These myths helped to propagate a moral geography in which rural places and rural people were located as the repositories of 'true national values' – and later informed a politics of exclusion in rural space with racist and xenophobic undertones (see chapter four). They have also been drawn upon by the Countryside Alliance and rural protesters who have reproduced their own modern myths of the countryside as a space of freedom and liberty, and of the countryside as a disempowered, beleaguered space.

These 'discourses of rurality' have long been entwined with discourses of power to promote particular hegemonic (or proto-hegemonic) representations of rural space and society that have served the interests of dominant power elites (Woods, 1997). As will be discussed at greater length in chapter three this purpose was most explicitly advanced in discourses such as those of the 'country gentleman' and of the 'agrarian community' which constructed power in terms of rurality, and rurality in terms of power. The idea of the 'apolitical countryside' was a similarly mythic construct, designed to direct attention away from the political structures that did exist in rural Britain and to discourage challenges to the existing power elite.

The principles of the discourse of the apolitical countryside are simple. Politics, it suggests, are a modern, frivolous invention produced by the alien and morally corrupt society of the city. Politics are divisive, setting class against class, party against party, and generating violent demonstrations and damaging industrial strikes. Moreover, politics encourages a fascination with the ephemeral and the trivial and works against the natural order of things. The countryside, in contrast – so the myth goes – is characterized by a stable and order society, by the natural leadership of a landed elite, and by natural, informal and consensual governance. The ordinary working people of rural areas are too busy engaged in virtuous and productive labour to have time for politics, and they identify more readily with their community – in which their interests can be elided with those of their employers and landlords – than with their class. In sum, urban society and politics were portrayed as artificial, in stark contrast with the naturalness of the rural.

Significantly, these ideas gained currency in the 1920s, 1930s and 1940s. They were reproduced through a variety of media, including the fashionable local pageants that were hosted by small towns across Britain in the first half of the twentieth century (see Woods, 1999), and the burgeoning literature of rural community studies. Notable among the latter is a study of the community of Luccombe, on Exmoor in south-west England, by W. J. Turner, published in 1947 as *Exmoor Village*. As Matless (1994) observes, Turner's account of Luccombe invested the village with a sense of permanence, 'a stability contrasting with, and holding lessons for, the changing world around' (p. 24). The community was explicitly and overtly represented as *different* to the city, and that difference included the absence of politics:

> The men of Luccombe all work with their hands – with horses, trees, stones and paint ... Life is therefore very much stripped of all superfluities, and most of the questions that are hotly debated in cities and big industrial centres have no interest whatever for Luccombe people as they have more serious business of their own to attend to. There are strictly speaking, therefore, no political opinions or discussions. (Turner, 1947, pp. 30-31).

Turner hence attributes the apparently apolitical current of country life to the nature of rural work and to the connection of rural people to the enduring harsh realities of everyday existence. In turn, these characteristics of the countryside are used to suggest that rural life is morally superior to urban life and to valorize the absence of politics, Matless (1994) for example noting that,

> The permanence of meaning resident in Luccombe was contrasted to what were regarded as the whims of fashion and politics, urban ephemera less enduring than the deeper reality of the country. Luccombe becomes a site of the real, a place of vital meaning rather than of the superficial language of the media and the politician. (p. 24).

However, the underlying biases of this representation are revealed in the two pages that Turner devotes to his chapter on 'politics'. Noting that there was only one declared Labour supporter in the village who would talk 'vigorously and dogmatically against the "gentry"', Turner comments that 'more typical' were the remarks of one villager about a visitor to Luccombe:

> He was Socialist, you know. Not the sort of man I'd like to be seed out with. You'd soon get into a barny, you know. He'd got no time for gentry nor the likes o' they. I mind well the day I went out with him to buy some cigarettes. He was running down the farmers all the time, and I was very glad to get back indoors. No, I don't say you'll find any Socialists among the farm-workers here. (p. 45).

What becomes clear here is that politics is being associated with Socialism and with change – with a threat to the existing order. The entrenched Conservatism of rural society and the established paternalistic power structure headed by the landed gentry and large farmers, in contrast, were represented as being 'natural' and therefore not political. In this way the propagation of the myth of the apolitical countryside helped to shore up the traditional power structure in rural areas at a time of great social and economic change and political uncertainty. As will be discussed in detail in chapter two, the leadership and authority of the old landed elite was seriously tested during the inter-war period as a result of economic pressures and the consequences of aristocratic deaths in the First World War. As the aristocracy and gentry retreated their place in the rural power structure was taken by the growing ranks of independent farmers – but as landowners and employers such farmers shared a broad ideological outlook with the gentry and the transition between dominant elites represented more the expansion of a hegemonic power bloc than a radical redistribution of power (Woods, 1997).

The rise of the labour movement nationally, however, posed a more dangerous challenge to the rural elite. Trade unions had organized among farm labourers and other rural workers and Labour party branches were beginning to be formed in rural towns and villages – often in spite of considerable intimidation. At the 1945 General Election, Labour captured over fifty rural or semi-rural constituencies, particularly in the Midlands and East Anglia. In constituencies such as Norfolk South West, where Sidney Dye, a local farmer and member of the National Union of Agricultural Workers, won with a slender majority of 53 votes, Labour's success depended on appealing to both farm workers and small farmers. It was precisely the prospect of this kind of alliance that motivated the reproduction of the discourse of the apolitical countryside which sought to warn off small farmers and the rural working class from flirting with Socialism, and encouraged them instead to identify with their rural community and thus to implicitly accept the leadership of the established elites.

In doing so, however, the discourse of the apolitical countryside showed blatant disregard for the distinguished history of radical politics and protest in rural Britain. From the fourteenth century Peasants Revolt to the Captain Swing Protests in southern England in 1830 and the Rebecca Riots of rural Wales a decade later – not to mention periodic food and rent riots and protests – the British countryside has been a space of discontent (Charlesworth, 1983; Mingay, 1989). It was in rural Surrey that the Diggers established their revolutionary commune in 1649, and it was in rural Dorset that the sentencing of Tolpuddle Martyrs in 1834 provided a watershed moment for trade unionism (Charlesworth, 1983; Parker, 2002). Even as the land-owning elite was promoting the idea of the apolitical countryside in the 1930s, farmers were demonstrating in London against the Labour government (Griffiths, 1999).

Moreover, by representing politics only as action that brought about *change*, the discourse of the apolitical countryside denied the politics that were intrinsic in the continuing stratification of rural society and the concentration of power with an enduring elite whose foundations dated back to the mediaeval period, and whose status was built on a political manipulation of rural identity and rural space.

The Foundations of Rural Power

One erroneous aspect about the construction of the rural as apolitical was that the perceived stability was deeply entrenched in a highly-developed political order. The degree of domination achieved by the landowning elite may have discouraged open political contest, but it itself involved a complex web of political relations.

Emerging from attempts after the Norman conquest to balance the centralizing tendencies of the new regime with the provincialism of Saxon England (Redlich and Hirst, 1958), the institutions of rural politics gradually evolved over eight centuries. Essentially these operated at two levels. At a community level, manorial administration was conducted through the court baron and the court leet, bodies that were supplanted by the parish vestry as the absolute power of the lord declined (Eastwood, 1994; Webb and Webb, 1963). At a county level, increasing centralization saw the creation of the offices of sheriff (11th century), justice of the peace (14th century) and lord lieutenant (16th century) – offices that are still in existence today (Lee, 1963; Packett, 1975).

At the same time there evolved from the feudal mediaeval lords a governing elite to occupy such offices. The basis of membership of this 'aristocracy' was complex but hinged on the hereditary principle and landownership, the latter being important not only because of agriculture, but because of the allegiance of tenants it secured and the cultural capital it bestowed (Girouard, 1978). Hence landowners formed an oligarchic elite. They dominated the Quarter Sessions, which by the 19th century 'increasingly looked like a nominated county Parliament' (Beckett, 1986, p. 393); whilst the lord of the manor balanced power with the incumbent minister and the other freeholders of the parish who formed the vestry, 'in no sense a body representative of the population as a whole' (Webb and Webb, 1963).

Attempts to reform the self-appointing oligarchy persisted throughout the 19th century. Poor Law Unions, established in 1834, took power from the parish vestries and were followed by health, burial, highways, sewerage, drainage and school boards (Newby *et al.*, 1978). In 1888 the Quarter Sessions were replaced by elected county councils, which were complemented by parish councils in 1894 (*ibid.*). However, the institutionalization of rural government 'did not appear to present any challenge to the traditional hegemony of landed interests' (*ibid.*: p. 225; see also Young and Mills, 1990).

The ability of the ruling elite to adapt to new government structures reflected the strength of the hegemony they had constructed. This was based on three foundations: firstly, particularistic power relations within individual parishes; secondly, the flow of patronage and influence at county level through the social networks of the landed elite; and thirdly through a cultural and ideological hegemonic discourse which presented the countryside as a close-knit but stratified agricultural community.

Parish Power

The transformations of the agrarian revolution had replaced the communal basis of agriculture within self-sufficient villages with an agricultural system based upon private landowning and contractual employment, on which the majority of rural residents were now economically dependent. The village was hence recreated as 'an 'occupational community', one whose whole existence was intimately bound up with the fortunes of a single industry – farming' (Newby, 1987, p. 77).

The consequences of this were two-fold and mutually reinforcing. The localized basis of the occupational community meant that the fundamental unit of political organization of rural areas was the spatial focus of that community - the parish, manor or estate. At the same time, the distribution of power within that (or over that) community was largely determined by the ownership of land.

Such was the closed nature of the rural village, and the central importance of land as a resource for both production and reproduction; the labouring class of the village became dependent on the landowner for both employment and accommodation. With almost total power within the economic and social spheres, landowners were virtually guaranteed political power within the parish. This was further reinforced by the nature of the organization of employment and tenancy, which were managed on an individual basis through particularistic power relations, thus making political organization amongst agricultural labourers difficult. (Newby *et al.*, 1978).

Thus as Newby (1987) describes, the 'great estates' of the aristocracy were the centre of political and social influence extending into the surrounding area:

> This involved a complex set of proprietorial rights, not only over the agriculture of the estate, but over its minerals, its game, its Members of Parliament, its clergy – in short over the entire locality and its inhabitants. (p. 61).

This control was paramount in the home estate villages of the 'greatest' aristocrats, but was reproduced to some degree in smaller estates of the minor aristocracy and gentry. In the peripheral parts of the larger estates, the personal dimension may have been weaker, but the power of the landowner remained, exercised on their behalf by their agent, or a trusted representative such as a tenant farmer or vicar (see also Howkins, 1991).

The particularistic power structure was however, to a large degree, dependent upon the 'closed' nature of the village. It therefore served the landowner's interest to perpetuate a localized ideology of 'community' and endear a strong identity and loyalty to the parish amongst residents (Saunders *et al.*, 1978). As Newby (1987) describes, landowners expended effort to cultivate an idealized 'organic community', in which a benevolent squirearchy, a respectful tenantry and deferential farm workers lived together in an ordered and stratified world. This cultivation of 'community' was deliberately oppositional to other villages in the area, and to 'the more impersonal ties of class' (*ibid.*, p. 89). It was also specifically spatial:

> The parish also represented a spatial framework within which these orderly and harmonious relationships could be played out - ideally a self-contained little enclave, cut off from the potentially subversive trends and ideas of the outside world ... this sense of community was also to be expressed visually and aesthetically in a landscape which conferred a sense of place ... the village should consist of clearly-defined 'natural orders' who rarely ventured beyond the parish boundaries – apart from their upper- and middle- class emissaries who were the representatives of the village to the outside world – and who lived in a picturesque setting of parkland, manor house, church tower and quaint cottages. (Newby 1987, p. 89).

The rural village community was thus, in Cohen's (1985) words, 'symbolically constructed'. The landowner would often go to great lengths to promote this construction, engaging popular culture and ritual such as annual festivals and activities such as hunting. Both contributed to what Cohen labels, 'the myth of egalitarianism'. In fact, 'the inhabitants of the Big House always remained outside the village community, both geographically and socially, cultivating an ideal of community rather than participating in the reality' (Newby 1987, p. 62).

Cohen argues that boundaries are of particular importance in the symbolic construction of community. He further contends that boundaries may be perceived differently by different members of the community and on different occasions, with, for example, festivals and carnivals providing opportunities through which boundaries could be contested or subverted on a temporary basis, thus contributing to their more permanent reaffirmation. Moreover the boundaries that constituted the symbolic community were both internal and external. A boundary existed within the village community between the workers and the residents of the 'Big House'. This was recognized and reinforced from both sides, but was not allowed to detract from the primary community of the village of the whole, and its external boundaries. The parish provided the geographical basis for this, yet its boundaries were more than just lines on a map, but were symbolic constructed and marked by the community's members in opposition to the outside through, for example, ritual and everyday practices of personal interaction, such as localized dialect and the use of nicknames (Phythian-Adams, 1989).

The symbolic importance of the parish was reproduced institutionally by Rural District Councils, which, as Stanyer (1975) observes, 'based their system of electoral areas on the principle of the representation of parishes rather than population' (p. 280). The councils were composed of single parish wards, although larger parishes may have had two or three members. This nevertheless still under-represented large villages and small towns, a system which 'tended to give an advantage to farmers and landowners' (*ibid.*). In estate villages the single parish wards virtually guaranteed the squire an almost *ex officio* seat on the RDC.

County Power

The Local Government Act of 1888 reformed the county government structure by transferring power from Quarter Sessions to elected county councils (Howkins, 1991; Lee, 1963). However, it is significant firstly that counties remained the basic administrative unit; and secondly, that the power elite at county level remained the same, surviving the political upheavals in tact. This can partly be explained in terms of resources. For instance, as Stanyer (1989) observes, transport and money were of primary importance: 'throughout the county there was a level of society which contained people – landowners in rural areas and capitalists in urban areas – who were able to make the day's journey necessary to serve on a county council ('government by horse and trap')' (p. 79).

The bias of resources was reflected in semi-formal requirements for some positions. Thus, the Lord Lieutenant was always chosen from a landed family as 'since they had to act as the Queen's hosts when she visited the county, they needed a house big enough to accommodate her and her entourage' (Paxman, 1991, p. 64). These requirements, however, merely reflected the success of the landed elite in perpetuating their own power, which was dependent upon the continuance of the county as the unit of government.

By the late nineteenth century, the gentry were themselves socially organized at county level – hence 'county society' and 'county families'. Whilst the more important aristocrats would be primarily a part of metropolitan society, they would return periodically to maintain their place in the county social network of the lesser aristocracy and gentry, through which the flows of power, patronage and influence which contributed significantly to the county's governance progressed. An outsider wishing to achieve power in a county's government needed first to integrate themselves into the gentry's social network (see chapter two).

Thus, as Guttsman (1963) observed, a landowner, coming into his inheritance as an English Country Gentleman could look forward to a career as 'a magistrate, guardian of the poor, Deputy Lieutenant of the County or Sheriff and later often as a member of the County Council' (p. 144). Indeed, as late as the re-organization of local government in 1974 there existed county aldermen who tended to be disproportionately drawn from the landed classes, and in many counties the council chair was almost invariably a landowner or senior military officer (*e.g.* see Stanyer, 1989; Newby *et al.*, 1978).

However, as with the localized parish power structures, for the power of the landed elite at county level to be noticed and accepted, it was required to socially reproduce the county as a space with meaning for its inhabitants, and to endear feelings of belonging and loyalty. As with parishes, this required the symbolic construction of community, but the county was an 'imagined community' (Anderson, 1991), as there would be no personal interaction between the majority of its members. They could, however, share in common boundaries and a common culture.

Hence 'county space' was socially produced and symbolically bounded through specific 'county' histories, traditions and folklore. An almost quasi-ethnic distinction between the residents of different counties was peddled in popular culture. Whilst intense loyalty was inspired by institutions such as cricket clubs and county regiments. Furthermore, loyalty to the county meant loyalty to the county elite as the production of a county identity often involved the positioning of the elite families as an integral part of its history and culture, for example through public drama, ceremony and publications (Woods, 1997, 1999).

The Rural Dimension

The traditional rural power structure was therefore framed with a distinct localist spatiality. Yet it also had a particular rural dimension. The importance of land and agriculture in structuring particularistic power relations and building village communities has already been noted. Land though also had a symbolic importance, supporting a social and political order based on aristocratic values (Newby, 1987). The symbolic importance of land depended upon the perpetuating of a belief in the countryside as an agricultural space – 'the pastoral myth' (Short, 1991). Hence, the political order in rural Britain was in turn dependent on the reproduction of this particular 'discourse of rurality'. Equally importantly, changes in the political balance of rural areas can be connected with the import of new ideas that have challenged the constructs of rurality held by the established elites. What all these discourses of rurality tend to share, however, is a desire to symbolically differentiate the countryside from urban areas.

In the early twentieth century, the conversion of these symbolic boundaries into spatial boundaries was increasingly sought. The Council for the Preservation of Rural England, for example, was formed as part of an impetus for the ordering of rural space, which included an ordering of the rural as separate to the urban (Matless, 1990, 1998). This movement had a particular political agenda in that as the landed classes had lost political control of the towns (Beckett, 1986), they needed to protect their power in rural areas by partitioning off the countryside as different.

The separation of the government of town and country had begun with the founding of boroughs in the mediaeval era. Here authority was exercised by the burgesses – the freeholders – who were often organized through craft guilds which were the real loci of power (see Phythian-Adams, 1972). Even in unincorporated towns, commercial freeholders gained some degree of influence in a delicate balance of power with the manorial lord and church augmented through the vestry and court leet. This trend was accelerated by municipal reforms in 1835, resulting not in a radical democratization, but in the empowering of a business elite. Thus the hegemony of the landowning elite in rural politics in the early 20th century was mirrored in a hegemony of an elite of businesspeople and professionals in small town politics (see Birch, 1959), closely associated with institutions such as the Chamber of Trade, Rotary Club or Masonic lodge. Equally these institutions played an important part in the symbolic construction of a localist community, expressed also through civic ritual, and the related presentation of that community as a 'market town' thus elevating the status of the town's traders.

The absence of such civic institutions in rural communities, in contrast, was taken to be a defining distinction between urban and rural society. Thus, whilst town dwellers were considered to be *citizens*, participating in the self-government of their cities in a contract-based society of exchange, country dwellers were regarded as subjects, *governed by* the local squire in a political system based on personal authority and interpersonal ties of employment and tenancy (Woods, 2003a). It was not until the establishment of county councils and parish councils that rural communities were granted the opportunity for effective self-government and, as noted above, the success of the established elites in colonizing the new councils meant that the democratization of rural society on a scale comparable with urban areas was in reality delayed well into the twentieth century.

The National Dimension

The feudal legacy shaped not only the exercise of power within rural areas but also the representation of rural areas and interests in national politics. Not only were the great aristocratic landowners entitled to sit in the House of Lords, but members of the House of Commons for the (rural) county divisions were also disproportionately drawn from the ranks of the aristocracy and landed gentry. According to the 1872 Survey of Land Ownership, there were over one hundred

MPs who were among the ten largest landowners in their home counties, including the Prime Minister, William Gladstone, who with 6,908 acres was the second largest landowner in Flintshire (Cahill, 2001). Where large landowners did not sit in parliament themselves their patronage frequently determined who did. Yet, more important than this dominance of formal political positions were the social networks of the aristocracy and landed gentry, organized around family ties, public schools, universities, regiments and London clubs, which provided the real conduits of rural power and influence at a national level (see also chapter two). This amounted to an effective privatization of rural government – rural areas were represented in national politics through the personal networks of large landowners who in turn were left to govern their own estates and localities as they chose.

It was arguably only with the development of industrial capitalism and the need to manage the exploitation of rural resources in the interests of capital accumulation that the state itself became directly involved in rural governance. A Board of Agriculture was established in 1889 – some 27 years after the creation of the Department of Agriculture in the United States – as a result of a Royal Commission on the Depressed State of the Agricultural Interest (Winter, 1996a), and became the Ministry of Agriculture in 1919. The Rural Development Commission was established in 1910 and the Forestry Commission in 1919. However, as Winter (1996a) observes, the staff of the early Board of Agriculture were 'recruited by patronage from the ranks of the landed elite' (p. 84), whilst the very constitution of the rural development and forestry bodies as 'commissions' indicated their paternalistic motivations. Thus, such institutional developments served to formalize the process of governance for rural Britain, but did little to disturb the rural balance of power.

The culture of accommodation continued with the incorporation of the National Farmers' Union and the Country Landowners Association into corporatist policy-making structures for agriculture after the First World War (Winter, 1996a). Hence, by the third decade of the twentieth century a division of labour had emerged in British rural politics that was to endure for the next seventy years – between a process of *external representation*, that was the domain of farm unions, business associations and pressure groups operating within exclusive policy communities, and a process of *internal governance*, dominated by land-owning and agricultural elites and their allies.

The subsequent development of both parts of this model is explored in late chapters of this book. Chapters two and three focuses on the transition of the internal power structure of rural areas, following the decline of the squirearchy, the rise of the agrarian elite and the coming of the service class. The development of the external mode of representation of rural interests, meanwhile, is examined in chapters four and five, which look at rural representation through political parties and pressure groups respectively.

As these chapters will demonstrate, there is across both the internal and external dimensions of rural politics a shared chronology that points to a fundamental re-formation in the late twentieth century. For most of the century the structures and processes of rural politics evolved in a manner that responded to changing circumstances but remained essentially true to the foundations of rural power outlined above. From the 1970s onwards, however, the intensity of social and economic restructuring in rural areas was sufficient to undermine the stability of the old order, introducing new actors into the rural political arena, creating new formations and shifting the locus of power. Even the content of rural politics has been transformed as the focus of political dispute and conflict shifted from resource management and service delivery to issues that revolve around the very meaning and regulation of 'rurality', arguably producing a new 'politics of the rural' (Woods, 2003a). The details of this transition are again discussed in later chapters, but the remainder of this chapter sets the context for this argument by outlining the key processes of rural restructuring and the major ways in which they have impacted on rural politics in Britain.

Rural Restructuring and its Political Impact

It has become an axiom of rural studies that the contemporary countryside has experienced (and is still in the throes of) widespread social and economic restructuring. Research has accumulated vast amounts of empirical evidence of change in rural Britain over the last three decades, from adjustments in the rural labour market to migration trends. Yet such accounts of rural restructuring need to take care not to reproduce a false dichotomy between a dynamic rural present and a stable rural past, but must rather recognize that the rural has always been a space of change and development and that from a long historical perspective many rural areas have experienced periods of considerable upheaval on at least an equivalent scale to the contemporary era at various points in the past (Woods, 2005a). In Britain such historical episodes include the process of enclosure during the first agricultural revolution, the Highland clearances in Scotland, and the period of rapid industrialization and urbanization in the nineteenth century, all of which had profound effects on rural society.

Moreover, Hoggart and Paniagua (2001) critique the loose usage of the term 'restructuring' in contemporary rural studies. They argue that whilst 'rural restructuring' can be employed with a fairly precise and theoretically grounded application, the notion of 'restructuring' is also frequently used to refer to relatively minor changes in a manner that denudes the term of its conceptual resonance. In arguing for a stricter understanding of 'restructuring', Hoggart and Paniagua suggest that 'rural restructuring' should both be multi-sectoral and be measurable quantitatively and qualitatively:

> For us, when seen as a shift in society from one condition to another, 'restructuring' should embody major qualitative, and not just quantitative, change in social structures and practices. Unless we want to trivialise the concept, its use should be restricted to transformations that are both inter-related and multi-dimensional in character; otherwise we have descriptors that are more than adequate, like industrialisation, local government reorganisation, electoral dealignment or growth in consumerism. To clarify, in our view restructuring is not a change in one 'sector' that has multiplier effects on other sectors. Restructuring involves fundamental

readjustments in a variety of spheres of life, where processes of change are causally linked. (Hoggart and Paniagua, 2001, p. 42).

In seeking to position the late twentieth century and early twenty-first century as a period of fundamental rural restructuring it is necessary therefore to firstly differentiate the contemporary experience from earlier episodes of change and secondly to demonstrate the integration and interconnectivity of the processes observed. Indeed, it can be argued that it is evidence of the latter that helps to answer the former. Contemporary rural change is distinguished by the *pace and persistence* of change – in that rural economies and societies are not just changing, but changing constantly and rapidly in response to successive trends and innovations – but also by the *totality and interconnectivity* of change (Woods, 2005a). Many historical instances of rural change, such as enclosure, were revolutionary for those directly affected but were spatially limited in their impact. In contrast, the contemporary countryside is subject to processes of change whose implications are manifest across numerous different sectors of economic and social activity and reach around the world.

Three dynamics in particular may be identified as the drivers of contemporary rural restructuring. Firstly, rapid and advanced technological innovation has reshaped both the organization of the rural economy and the social practices of rural life. The development of motor transport, for instance, assisted the mechanization of agriculture, permitted food, timber and other natural resource products to be transported over longer distances, and enabled individual rural residents to travel more freely for work, leisure and shopping, loosening ties with the community of residence and facilitating the growth of commuting. Telecommunications technologies, meanwhile, have both alleviated some of the disadvantages of peripherality, enabling rural localities to compete with urban locations in new economic sectors, and allowed rural residents to consume the same cultural commodities and experiences as urban residents, diminishing the significance of localized rural traditions, events and cultural practices.

Secondly, technological 'modernization' has been accompanied by 'social modernization'. This includes the development of mass participation in further and higher education, increasing social mobility by broadening the range of employment opportunities available to individuals and encouraging greater mobility out of rural communities for training and employment; the decline of organized religion, which had once been a core tenet of rural community life, along with a wider shift in popular values that has challenged aspects of the established rural social order; new ways of using leisure time, which have promoted the growth of rural recreation and tourism; the rise of environmentalism, and with it new ways of valuing rural landscapes; and the advent of a consumer-oriented society, in which our increasingly sophisticated (or cost-conscious) tastes in food and other rurally-sourced goods have created new demands on producers, and in which individuals have brought lifestyle preferences into their residential decisions, fuelling the trend of counterurbanization.

Thirdly, the above two dynamics have been entwined with a spatial reconfiguration that can be represented as a state of advanced globalization in which rural localities are locked into a matrix of global interconnection and interdependence. Within this, rural areas have been particularly affected by the globalization of trade (including, notably, agricultural trade), and the accompanying concentration of power with transnational corporations and regulatory bodies; by the globalization of mobility, including the global-scale movement of tourists and of migrant workers; and the globalization of values, through which rural traditions such as hunting and local lay knowledges of nature have been challenged by the promotion of new sets of universal rights and ethics (Woods, 2005a).

Evidence of Rural Restructuring in Britain

The impact of the above processes of restructuring in rural Britain can be illustrated by highlighting the quantitative evidence for four of the most notable trends: the readjustment of the rural economy, including the decline of farming as a source of employment; the growth of commuting; counterurbanization; and the social recomposition of the rural population.

In 1950, over a third (34.6 per cent) of the 'rural population' of Britain was estimated to be dependent on agriculture for its income. By 1970 the proportion had fallen to 24.3 per cent, by 1990 to 19.6 per cent and by 2000 to 16.8 per cent (Woods, 2005a). Not only did agricultural employment decline, but the nature of farm employment also changed. The number of hired farmworkers fell sharply from over 800,000 in the 1940s to under 300,000 in the 1990s (Clark, 1991). Whereas there had been nearly three hired farmworkers for every farmer in 1931, by 1987 the ratio was 1.1:1. Other 'traditional' rural economic activities based on resource exploitation, such as forestry and quarrying, also declined in significance, as did manufacturing, which had been the largest employer in non-metropolitan areas in 1951. As Table 1.1 shows – for the slightly different geographical base of non-metropolitan areas – decreasing employment in these areas was balanced by increased service sector employment. Furthermore, whilst agricultural employment was usually undertaken in or close to a worker's home community, the restructuring of the rural economy was accompanied by a 25 per cent growth in commuting in significantly rural regions between 1980 and 1990 (Schindegger and Krajasits, 1997).

The expansion of commuting is a factor in counterurbanization, or the trend of migration from urban to rural localities. Population growth in rural Britain outpaced that in urban areas throughout the 1970s, 1980s and 1990s. Between 1971 and 1981 the population of non-metropolitan counties in Britain increased by 6 per cent whilst that of metropolitan counties decreased by 6.5 per cent (Serow, 1991). Similarly, a 12.4 per cent increase in the population of rural districts in England between 1981 and 2001 compares with a population increase of only 2.4 per cent in urban districts over the same period (see Table 1.2). These aggregate figures reflect the cumulative effect of a complex set of migration dynamics between and within urban and rural localities, as well as natural changes to the population, yet the strength of direct urban to rural migration is indicated by the 100,000 people who were recorded as moving from urban districts to rural districts in the year preceding the 1981 census (Serow, 1991).

There are also, however, differences in the migration trends for different sections of the population, with a net balance of out-migration from rural areas by young people aged between 18 and 30 compared with a strong current of retirement migration to the countryside. This has contributed to a demographic recomposition of rural communities which have become distinctly more elderly in character when compared with urban areas. In 2001, for example, 37.3 per cent of the population in rural England was aged between 15 and 44, compared with 43.2 per cent of the urban population, whilst 18.1 per cent was aged over 65, compared with 15.0 per cent of the urban population (Countryside Agency, 2003). Similarly, there has been a recomposition of the class structure of many rural communities. It has been estimated that around 40 per cent of in-migrants to rural areas between 1970 and 1988 were members of the 'service class' of professional and managers (Halfacree, 1992), whilst the effect of rapidly inflating property prices has been to provoke a undercurrent of working class out-migration.

Table 1.1 Employment in non-metropolitan districts* of England and Wales, 1951 and 1991

Sector	1951	1991	Change
Agriculture, forestry & fishing	9%	3%	-6
Manufacturing	32%	19%	-13
Services	27%	38%	+11
Distribution	11%	21%	+10
Mining and quarrying	7%	3%	-4
Construction	7%	8%	+1
Transport	6%	6%	=
Energy and water	1%	2%	+1

* Defined as all of England and Wales excluding the seven major conurbations and twelve largest free-standing cities.

Source: Census of Population, 1951 and 1991.

Table 1.2 Population change in rural and urban districts of England

	1981-1991	1991-2001	1981-2001
Rural districts	+7.1%	+4.9%	+12.4%
Urban districts	+1.4%	+0.9%	+2.4%
England total	+3.0%	+2.0%	+5.0%

Source: Adapted from Countryside Agency, *The State of the Countryside 2003*.

The Political Impact of Rural Restructuring

The currents of restructuring in rural Britain have impacted on both the internal and external dynamics of rural politics and power in five key ways (see also Woods, 2003a). Firstly, the significance of agriculture and natural resource exploitation to the British economy has been transformed by globalization, technological advances and the transition from a Fordist to a post-Fordist regime of economic regulation. Accordingly, the state's interest in supporting agrarian capitalism has diminished. From being an essential tool for post-war reconstruction, the programme of government subsidies for agricultural production has become a liability and a drain on public expenditure. As both successive British governments and the European Commission have struggled to reform agricultural policy the impact has been to shift the balance of priorities in the external government of rural Britain and to undermine the operation of the old policy communities, thus weakening the influence of formerly 'insider' pressure groups such as the National Farmers Union. At the same time, in attempting to develop a new strategy for the governance of rural Britain, the national state has adopted a new mode of 'governmentality' in which the countryside is re-imagined as an amalgam of communities through which residents can be engaged in their own self-government by meeting certain responsibilities. This strategy of 'governing through communities' was most notably articulated in the first Rural White Paper in 1995, but has been reproduced in subsequent policy developments (Edwards et al., 2003; Murdoch, 1997; Woods, 2003a).

Secondly, traditional power structures have been eroded from within rural society as a result of both economic and social restructuring. The decrease in agricultural employment, for example, has weakened the influence that large farmers once enjoyed due to their power over employees and tenants in a paternalistic labour system (Newby, 1977; Newby et al., 1978). Meanwhile, counterurbanization has contributed to the emergence of a more mobile and more educated rural population which has few ties to agriculture, and which has become increasingly reluctant to accept the authority of the established elites (Woods, 1997).

Thirdly, as the commodification of the rural landscape, culture and lifestyle has become more important to the rural economy than the physical exploitation of rural resources, so the dynamics of political mobilization in rural areas has changed. The commodification of the rural includes not just the expansion of tourism, but also investment in rural areas through counterurbanization and gentrification by in-migrants pursuing the mythological 'rural idyll'. As frictions emerged between the dream of the 'rural idyll' and the realities of rural living, in-migrants have been mobilized to act to protect their financial and emotional investment by opposing developments and activities that threaten the perceived 'rurality' of their home (Woods, 1997; Woods, 1998a).

Fourthly, the ability of established rural pressure groups to deliver to their constituencies has been undermined as the old policy communities have disintegrated producing dissent over their representational role. In response new, more militant, groups have emerged, many of which are prepared to engage in direct action tactics to advance their case. This was most powerfully demonstrated in September 2000 when a blockade of fuel depots initiated by Farmers for Action – a breakaway group critical of the National Farmers Union – brought much of the country's transport system to a halt (Doherty *et al.*, 2003; Woods, 2004). Other examples include the Rural Rebels in Scotland and the pro-hunting Countryside Action Network and 'Real Countryside Alliance' both of which are splinter-groups from the Countryside Alliance, itself an organization formed in the 1990s in response to the perceived failure of established lobby groups.

Fifthly, counterurbanization, the decline of agriculture and cultural homogenization have all contributed to a growing sense of beleaguerment amongst that section of the population that strongly identifies itself with traditional rural activities and values. This has been fuelled by the perceived lack of understanding of rural affairs by government at the national level and the bungled attempts of government to deal with crises including the collapse of consumer confidence in beef following the disclosure of the connection between BSE and CJD in 1996, and the foot and mouth epidemic in 2001. The notion of a 'rural-urban divide' has assisted the growth of the Countryside Alliance and mobilized participation in the Countryside Rally, Countryside March and the Liberty and Livelihood March (Woods, 2004).

Collectively, these processes have been part of a transition from 'rural politics' to a 'politics of the rural'. Whereas the former may be defined as politics located in rural space, or relating to rural issues, the latter is defined by the centrality of the meaning and regulation of rurality as the primary focus of conflict and debate. The potential for political conflict to be generated by the fragmentation of meaning of rural space was first detailed by the Belgian sociologist Marc Mormont in the 1980s. Mormont (1990) argued that restructuring had produced a situation in which there is no longer a single, homogeneous, rural space, but rather there are many different rurals existing in the same space. The inevitable tensions that can arise involve both local and ex-rural actors and often focus on issues of land development and spatial regulation. Thus, as Mormont comments:

> The structural characteristics of the development of rural space hide a series of economic, social and political contradictions which make up the basis of 'rural' forms of opposition ... From now on, if what could be termed a rural question exists it no longer concerns issues of agriculture or of a particular aspect of living conditions in a rural environment, but questions concerning the specific functions of rural space and the type of development to encourage within it. (Mormont, 1987, p. 562).

In the rural conflicts, or as Mormont labels them, 'rural struggles', that emerge as a consequence of this question, the central issue is 'the symbolic battle over rurality – that is, the legitimate definition of the term "rural"' (Mormont, 1990, p. 35). The examples discussed by Mormont, concerning tourism, access to public services and the right to local housing and employment, are all clearly located in rural space yet they engage actors from outside of the rural as well as mobilizing local actors. Thus we can identify two related scalar dimensions to the politics of the rural. On the one hand, Mormont observes that the 'rural struggles' he describes can provide the motivation for the emergence of new local political movements and actors within rural space:

> It is significant that ... rural life tends to restructure itself – at least partly – around the issues that have been raised in these struggles. Conflicts around rural schools, tourist developments and so on result in the setting up of new local associations which then involve themselves in plurilocal movements. There is hence a truly political effect, made as certain strata of the population thus give themselves means ... of expressing their demands, of organizing mobilization around issues which had, until then, been monopolized by the traditional political organizations. (Mormont, 1987, p. 566).

As will be demonstrated further in later chapters, this effect has been widespread in rural Britain over the past quarter of a century as conflicts that may have been sparked by seemingly minor disputes over access to footpaths or streetlighting as well as by larger-scale development and service provision issues, have mobilized new political actors who have organized petitions, mounted demonstrations, stood for election to parish, district and county councils challenging established councillors, and, in same cases, formed new pressure groups or political associations.

On the other hand, few rural conflicts of this type have been wholly contained within a rural locality. Conflicts may arise because of the intervention of an external actor, or local activists may discover that disputes cannot be resolved at a local scale, but must be taken to a higher authority. This is more true in Britain than in many countries given the highly centralized nature of British government and the relatively limited autonomy enjoyed by local government, for example over planning policy. As such, local campaign groups evolve to form coalitions with similar organizations in other localities and develop networks that extend up from the local to regional and national scales of articulation, enrolling new actors as they progress. In this way rural issues have begun to impinge on the national political consciousness.

Cox (1998) describes these two scalar dimensions as the 'space of dependence' and the 'space of engagement'. He argues that the space of dependence reflects the mobilization of political actors at a local scale because of their commitment to a particular space or place, and the way in which their legitimacy to act follows from their local identification. Yet, he argues, in order to achieve their objectives, local political actors frequently operate within a 'space of engagement' that transcends the local scale to connect with actors at regional and national scales, albeit within a fairly tightly defined sphere of interests. Significantly, Cox illustrates his thesis with reference to work by Murdoch and Marsden (1995) that examines a conflict over gravel extraction in rural Buckinghamshire. In the case study, proposals to open a gravel extraction pit near the village of Chackmore provoked considerable local opposition and the formation of a campaign group, Chackmore Against Gravel Extraction (CAGE). The campaign was motivated by concerns about the impact of the development on the local area, including despoilment of the landscape, increased traffic and pollution – issues that defined its space of dependence. However, the developed could not be successfully opposed by purely local action. The proposal was part of a regional strategy, formulated by ex-local actors, and followed a rationale of spreading the concentration of mineral sites. The campaigners hence had to engage with actors outside the locality. The space of engagement was therefore partly defined by these actors who were responsible for the proposal, but in response CAGE constructed its own network to enrol actors at the regional and national scales. This was done by shifting the focus of the campaign from the disturbance to local people to the potential impact of the mineral extraction in lowering the water table which in turn would effect the ornamental lakes in nearby Stowe Park, a site which could be represented as a nationally important landscape. As a result the campaign was able to enrol into its network the National Trust charity, which owned the gardens, and former pupils of the public school that occupied Stowe Hall. Collectively these actors provided an authorative voice that was accepted as legitimate at a national scale, and direct access to national policy-makers. By mobilizing this extended network the campaign successfully prevented the development of the minerals site.

Developing this thesis, Cox argues that local politics cannot be defined as politics that occur at a local scale but that local politics can appear 'as metropolitan, regional, national or even international as different organizations try to secure those networks of associations through which respective projects can be realized' (Cox, 1998, p. 19). Extending this logic further, it equally becomes possible to argue that 'rural politics' cannot be defined as politics that occur within rural space, but that a politics of the rural transcends rural and urban space enrolling actors in diverse locations and at a range of scales.

Conclusion

The politics of the British countryside are far more dynamic and complex than is often imagined. In spite of popular perceptions of the traditional countryside as an apolitical space, devoid of conflict and united in organic cohesiveness, rural society has always been highly politicized insofar that it was organized around a highly elitist power structure involving the parallel dimensions of internal power maintained through paternalism and external representation exercised through formal, top-down, organizations operating within exclusive policy communities. What was true was that these structures became naturalized, not least through the reproduction of the discourse of the apolitical countryside, such that the belief was popularly held that all classes with rural society shared common interests that rural leadership was social rather than political. This discursive settlement left little room for active political debate and the lack of visible conflict was wrongly read as an absence of politics rather than as an expression of a hegemonic paternalistic system.

In the latter part of the twentieth century, however, this old order was undermined by the effects of social and economic restructuring in rural space. New political actors emerged to challenge the leadership of the established elites, often mobilizing around issues that concerned the regulation or development of rural space. Such political actors were frequently frustrated to find that issues could not be satisfactorily resolved at a local scale, and were hence forced to constructed spaces of engagement that reached up to the regional and national scales of politics. This, combined with the collapse of the policy communities in agriculture and other rural policy fields, contributed to the emergence of rural issues on to the national political agenda, as countryside protestors have staged number of well-attended demonstrations, as disgruntled farmers have blockaded fuel depots and dairies, and as parliamentary time has been consumed by debates about hunting, public access to the countryside, rural housing development and the foot and mouth crisis. It is in this way that there has been an unexpected political awakening of rural Britain, articulated by a new 'politics of the rural' in which the very meaning and character of the countryside lies at the heart of the struggle.

This book follows the evolution of the 'politics of the rural' as it has gained momentum and jumped scales. The first part of the book examines the reconfiguration of local politics in rural areas. Chapter two builds on the historical survey outlined in this chapter by discussing the changing balance of local power in the countryside, including the decline of the squirearchy, the rise of the agrarian elite in the mid twentieth century and the more recent challenge from middle class in-migrants at the end of the century. Chapter three continues the narrative by examining the contemporary power structure in rural Britain, illustrated through a case study of Somerset.

Chapter four provides a link between the local and national scales. It discusses the historic connections between Conservatism and rural representation before arguing that a weakening of this relationship in the 1980s and 1990s contributed to the activation of a more direct form of rural representation, including the activities of the Countryside Alliance and other protest groups, which are explored in chapter five. The final two chapters each explore a major issue in contemporary rural politics. Chapter six concentrates on agricultural politics, including the responses to the BSE and Foot and Mouth epidemics and the rise of farmer militancy. Chapter seven, meanwhile, looks at conflicts over development in rural areas, particularly house-building, but also, briefly, road-building and the construction of windfarms.

Methodological Note

Chapters two, three and the first part of chapter four are all largely based on research conducted in Somerset between 1994 and 1996. The material presented here is drawn from a range of research methods including questionnaire surveys, semi-structured interviews, non-participant observation and the collation of information from publicly available sources including local newspapers, election literature, local history publications and directories and other documents deposited in local libraries. The questionnaire survey was the mail tool used for obtaining data about local leaders that could be employed in identifying elite networks. As such it asked questions about their record of involvement in local politics and government, employment and membership of social and political organizations. Because of the private nature of this information, no names were written on the questionnaires and each respondent was identifed only by a code number on the return envelope provided. The individuals to which questionnaires were sent were selected from information collected from published sources and included all serving county and district councillors in the Taunton Deane and West Somerset districts and all councillors who were defeated or retired at the most recent elections. Other questionnaires were sent to senior council officers; to appointed members of the health authority, NHS Trust and Exmoor National Park Authority; to officers of the constituency political parties; and to other individuals who were judged to be potentially influential in the locality including senior magistrates, leaders of local interest groups, former senior councillors and people who featured regularly in local news reports. A total of 240 questionnaires were mailed to individuals between September 1994 and January 1995, with a further 18 questionnaires sent to newly elected councillors in May 1995. Of these, 172 were returned with answers, a response rate of 66.7 per cent. These included 89 per cent of county councillors from the two districts, 77 per cent of West Somerset District Councillors, 68 per cent of Taunton Deane Borough Councillors and 81 per cent of appointed members of local governance bodies. The lowest rates of response were from business leaders, party officers and council officers.

The semi-structured interviews were conducted with 26 individuals between November 1994 and September 1995, including 15 serving councillors. The interviewees were mostly selected from the questionnaire responses to provide for both a range of experiences and the further exploration of particular issues. In additional to local politicians and other elite members, a small number of interviews were undertaken relating to key local issues, such as hunting, and with two local 'observers' – the editor of a local newspaper and a local historian. Where quotes have been used from these interviews they are identified by a code number for the interviewee in the form I000. Quotes taken from the questionnaire are similar identified by a code number in the from Q000.

The second part of chapter four and chapters five, six and seven are based on research undertaken between July 1996 and August 2004. The material presented here largely draws on information collated from published sources including newspaper articles, reports, documents and campaigning literature and websites. This document-based research has been complemented by a number of interviews with key individuals (who are identified by position when quoted), and by observation at a number of events including the Countryside Rally in 1998 and the Liberty and Livelihood March in 2002. In addition, information has also been drawn in a non-attributable way from informal conversations with national and local politicians, interest group officers and other activists in rural politics.

Chapter 2

The Changing Balance of Local Power in the Countryside

Introduction

The political landscape of rural Britain in 1900 appeared much the same as it had fifty, a hundred or two hundred years earlier. In spite of the social and political reforms of the nineteenth century, and the tide of rapid urbanization and industrialization, the quasi-feudal estate system described in the last chapter still dominated large swathes of the countryside. Within these territories the conduct of government continued to be effectively privatized and those who wielded power and occupied the offices of local leadership commonly bore family names that echoed down the centuries. Even the introduction of formal, elected, local government institutions, most notably county councils in 1888 and parish councils in 1894 had done little immediately change the balance of power. The creation of these institutions had initially been heralded as a radical step. *Punch* magazine published a cartoon entitled 'the great transformation' which showed a harlequin summoning the county councils, portrayed as a young maiden, into existence as the outgoing quarter sessions were represented as a cowering old fool. In Wales, Tom Ellis MP described the creation of parish councils a 'an instrument for uplifting the poor, for brightening the life and lot of the worker, and for increasing the agencies of a larger and humaner life' (Bowen Rees, 1994). The first elections to these bodies did mark a small degree of democratization. In Hawarden, Flintshire, 90 candidates contested 15 seats on the new parish council; the local squire failed to get elected to the parish council at Troed-y-aur, Cardiganshire; whilst councillors from commercial and professional backgrounds formed a majority on the new Kent County Council (Moylan, 1978). However, many of the small number of working class candidates that dared to contest council seats were reprimanded by their employers or landlords, and overall the aristocracy and gentry remained firmly in control. As Cannadine observed:

> Throughout the country as a whole, slightly over one-half of all newly elected county councillors were magistrates, and two-thirds of the counties elected the chairman of quarter sessions or the Lord Lieutenant as the chairman of the new county council. Many of the seats were not contested; most gentry who stood were elected; those few defeated were usually co-opted as aldermen; and so were those who did not even deign to stand. The first act of the Shropshire County Council was to appoint the Lord Lieutenant as an alderman. The Worcestershire County Council did exactly as Lord Beauchamp instructed it, even electing his complete slate of aldermen. At Stafford, Lord Harrowby greeted his fellow councillors with positively

patrician friendliness, including one man who turned out to be the porter. (Cannadine, 1992, pp. 158-159).

Yet, something had changed, however imperceptible it seemed at first. Cannadine notes that a 'democratic principle' had been introduced into rural government and that 'the hereditary and oligarchic principle had been discarded' (Cannadine, 1992, p. 157). The new councils at least allowed the possibility that individuals other than the landed classes might assume positions of local leadership in rural areas, in part by providing an independent structure for local administration that did not require leaders draw on their own resources to rule by personal largesse.

Elsewhere, other clouds were gathering that would soon disturb the Indian summer of the squirearchy. The issue of land reform, including land taxation and even the nationalization of land, was high on the political agenda and attracting support in the more radical parts of the Liberal party (Douglas, 1976; Vogel, 1989). Agricultural workers and tenant farmers were both starting to organize themselves into representative unions, particularly in Norfolk and Lincolnshire. Less obviously, but arguably more ominously, the geopolitical frictions of continental Europe were beginning on the intractable march towards a war that would eventually cost the British landed classes dear.

This chapter traces the evolution of the local power structure of rural Britain through the twentieth century. From the closing years of aristocratic dominance it charts the rise of the agrarian and small-town business elites and the more recent challenge that these have faced from new middle class elite fractions. In particular, the discussion focuses on three eras – the 1900s before the advent of the First World War, the inter-war period of the 1920s and 1930s, and the 1970s – and is primarily illustrated through a case study of Somerset, based on empirical research undertaken in the mid 1990s involving archival study and interviews.

The Dimensions of Rural Power

The ability of the squirearchy to extend their dominance of rural Britain into the twentieth century and the unraveling of their position after the First World War can both be explained by understanding the multi-dimensional nature of power in rural society. In this analysis it is contended that leadership within rural societies does not result from any single source but is based on a combination of three dimensions of power: resource power, associational power and discursive power.

Resource power may be described as the control over, or access to, specific resources enjoyed by an actor that enables that actor to do things that others cannot. This is a fairly conventional understanding of power in a broadly Weberian tradition that sees power as following from material bases. Thus, the 'resources' that endow this type of power include forms of material capital including money and property. However, resource power can also concern non-material resources, including forms of personal resources such as time, charisma, organizational skills, communication ability and so on (see Etzioni-Halevy, 1993; Woods, 1998b).

Rarely, though, does one individual or institution possess sufficient resource power in their own right to be able to exercise power at a societal level. Wealth on its own, for example, does not result in political authority. Instead, networks and alliances have to be created between actors with access to, or control over, different resources that can collectively create a 'capacity to act'.

Thus, *associational power* is power – understood as an ability to achieve things – that results from the blending together of a range of resources through a network or coalition of actors. It is power that comes from association. The significance of associational power was highlighted in actor network theory, as articulated in Latour's (1986) observation that 'when you simply have power – *in potentia* – nothing happens and you are powerless; when you exert power – *in actu* – others are performing the action and not you... History is full of people who, because they believed social scientists and deemed power to be something you can possess and capitalize, gave orders no one obeyed!' (pp. 264-5). Associational power is hence something that can only be attributed to a collective, but it should also be noted that individual actors may possess of resources of social interaction that make them particularly valuable in building networks of associational power. Such individuals are key actors in the construction of semi-permanent elite networks or regimes that serve to stabilize associational power by connecting actors with valued resource power in high-quality relationships (creating 'preemptive power' to use Stone's (1988) terminology) (see also Woods, 1998b). Stability can also be achieved by shrouding an elite with a cloak of legitimacy through the exercise of discursive power.

Discursive power refers to the ability of a person or an institution to act in a certain way because others believe that they are 'powerful' and accept their actions. Discursive power can therefore reinforce associational power by presenting a network of actors drawn together for the purposes of creating a capacity to act as a legitimate and natural leadership group. It can also prolong the leadership of specific elites or individuals by disguising the loss of resource power. In essence, discursive power results from the production, reproduction and circulation of 'discourses of power' – popularly-diffused beliefs and prejudices which establish the qualities expected of leaders, which present the elite as being worthy of power, and which define the power or influence that an elite might reasonably be expected to have. In other words, discourses of power create the circumstances in which a non-elite attributes power to an elite thus making them powerful (Woods, 1997).

The effect of discourses of power can be two-fold. Firstly, they establish the criteria against which potential leaders will be judged by presenting certain qualities as being more desirable than others. Gaining political office hence becomes in part a question of having the correct 'cultural capital' – a standard that has changed over time and thus can be read as part of the explanation for shifts in power. Secondly, discourses set the limits to political debate. They establish conventions about what are acceptable opinions to express or actions to take, about which subjects are the concern of local governance and which are not, about what topics or events are constructed as 'political', and about which arguments carry more validity than others. The discourse of the 'apolitical countryside' discussed in the previous chapter remains the most notable example of this in a rural context, illustrating the way in which discourses of power have frequently become entwined with discourses of rurality.

Discourses of power and discourses of rurality should not, however, be seen as ideologies around which parties are formed, from which policies are prescribed, and between which voters choose in elections. Although a number of discourses are labeled and discussed in this chapter, in reality none have been entirely coherent or discrete. Individual actors construct their own ideas about power, place and politics, which may borrow from a number of different discourses. Rather, the importance of discourses is that they provide a structure that informs individual action, and these 'big stories' can be influential in determining the overall direction of local politics in any given place at any given time, used by individuals as 'folk models' that help to define an individual's commonsense view of their world but differently received by different individuals, depending on their social and spatial context (Thrift, 1996). It is with this note of caution in mind that the next section proceeds to explore the role of the different dimensions of power in supporting the leadership of the squirearchy in early twentieth century Somerset.

The Twilight of the Squirearchy

At the start of the twentieth century, Somerset stood on the cusp of social and political change. The county had been spared large-scale industrialization or urbanization and remained largely rural in character, but manufacturing had established a presence in Taunton, Bridgwater, Street, Yeovil and Frome, and these towns, along with the seaside resorts of Clevedon, Burnham and Weston-super-Mare, had expanded rapidly. The new elected county council had provided the opportunity for members of urban professional and commercial classes of these towns to become involved in the local government not only of their own boroughs but of the county as a whole. Yet, as with most of rural Britain, such exploratory incursions only served to highlight the continuing dominance of a landed elite.

Of the 55 members of Somerset County Council in 1906 for whom an occupation or position can be identified (out of a total of 67 councillors), 26 were landowners, whilst 17 were drawn from commercial or professional backgrounds including two medical doctors, two auctioneers, a builder, a wine merchant, an insurance broker and a carriage manufacturer. The balance was held by eight farmers and three clergymen as well as one councillor described only as the local secretary of the Liberal Party. In the more senior political positions the grip of the squirearchy was even firmer. The 22 appointed County Aldermen included 15 landowners, 14 of whom occupied what *Kelly's Directory* described as 'principal seats', and four of whom had titles. Five of the county's seven Members of Parliament were landowners, whilst the other two, a banker and a solicitor, both came from upper class backgrounds.

Furthermore, the concentration of power was intensified by the multiple office-holding of several key individuals. The Chairman of the County Council, Henry Hobhouse, was also the Liberal Unionist MP for East Somerset, the constituency including his family seat at Woolston House. A barrister educated at Eton and Oxford, Hobhouse was described in one directory of 'Somersetshire leaders' as being 'of that type of public spirited English gentleman who are willing to give of their time and talent to the common weal without any reward' (Press, 1894). He continued to serve as county council chairman until 1924.

Across the county, Sir Alexander Acland-Hood Bt. was not only the MP for West Somerset (as his father had also been) but also a county councillor, magistrate and Deputy Lieutenant. The county's eighth largest landowner, he possessed some 11,000 acres around his seat at St. Audries on the Bristol Channel coast, the name of which he took as his title when ennobled in 1911. The combination of politics and land was regarded favourably by local commentators of the time, with one noting that 'his power and influence in the county [is] materially increased by the fact he is a rich, extensive landowner' (Press, 1894).

The leadership position of Hobhouse, Acland-Hood and their fellow landowners rested on the three dimensions of power discussed earlier. As the quote above implies, wealth was an important element of resource power, enabling landowners to participate in what was still the amateur, unrewarded, pursuit of local government. Equally significantly, freedom from the necessity of employment gave landowners time for administrative activities, whilst money also permitted access to transport. In an era before the arrival of motorized vehicles, and in a rural county where the penetration of railways was far from comprehensive, the ability to attend meetings of the council or the bench was largely limited to those with private transport, giving rise to what Stanyer (1989), writing about neighbouring Devon, calls 'government by horse and trap' (p. 79). It was land, though, that was the most important basis of resource power. Girouard (1978) suggests that land was significant to the aristocracy and gentry was not for agricultural production, but for 'the tenants and rent that came with it' (p. 2). By 1900, it is arguable that land was valued even more for another reason – discursive power. Crucially, it was ownership of land that enabled an aspiring local leader to comply with the discourse of the 'country gentleman'.

The Discourse of the 'Country Gentleman'

The social upheavals of the late nineteenth century meant that by 1900 the identification of a landed elite in rural Britain was no longer axiomatic and that an effort was needed for the elite to publicly define itself. In Somerset this was achieved in part through the publication of a number of books containing the biographies of 'Somersetshire Leaders' (Anon, 1908; Gaskell, 1906; Press, 1890; Press, 1894). Whilst the four volumes varied in their selection of 'leaders' and differed in their political biases, collectively the hagiographies they contained served to reinforce a distinctive discourse of power, entwined with an equally distinctive discourse of rurality.

The elite's discourse of rurality emphasized the importance of tradition, and the stability of rural society. It celebrated the role of the 'country gentleman' in the stewardship of nature – a concept that implied the management of both land and people – upholding the 'natural order' of hierarchies within both nature and human society. The discourse drew on the philosophy of Hume and Locke in constructing an ideology of property, as Rose *et al.* (1976) show; but also on the political ideology prevalent nationally a century earlier which Everett (1994) labels 'The Tory View of Landscape'. This conceptualization justified a hierarchical structuring of society, with wealth and status accompanied by obligations and duties, and a responsibility to the past as well as the future. These motifs are echoed in a passage from Gaskell's *Somersetshire Leaders: Social and Political* (1906), which epitomizes the elite's discourse of rurality:

> In the estimation of the majority of men there is no life that offers more solid attractions than that of a country gentleman. Far removed 'from the madding crowd's ignoble strife', he lives the pure natural life for which man was originally intended, and sits at the feet of the greatest of Gamaliels – Nature. Such a man is spared the bustle and turmoil of strenuous city life, with all its disappointments, its morbid restlessness and strifes. With him life pursues the even tenor of its way flowing with deep, silent stream, the converse of the career of a dweller in towns whose existence resembles rather a shallow, turbulent mountain torrent. But the life of the country gentleman doe not lack responsibility, for he owes duties both to his own position and to his tenants and dependants (no pagination).

The figure of the 'country gentlemen' hence blended together a discourse of rurality and a discourse of power in a manner that became expressed in the language of 'duties' and 'responsibilities' but which was unquestionably about privilege. As Newby (1987) observes, 'the gentlemanly ethic was thus a device for perpetuating distinctions of status, and an agency of social discipline.... The notion of gentility both conferred legitimate authority on those who held power in rural society and defended the rituals of social intercourse which accompanied the exercise of power' (p. 69). The 'country gentleman' would be accepted into the social and political elite, acknowledged by his workers and tenants as their 'better', and 'granted a legitimate right to rule' (*ibid.*).

Moreover, the discourse of the country gentleman was multi-layered, with each layer establishing qualifications for elite membership. Firstly, it was intrinsically *patriarchal*. Gentility was inherited through the male line and was based on a quasi-mythological masculinist ideal of the medieval knight. Strongly influenced by Arthurian legend (which had a particular resonance in Somerset), the gentlemanly ethic was identified with chivalry, such that those born to gentility were suggested to be 'naturally' attributed with the qualities of bravery, loyalty, fidelity, generosity and good judgment (Girouard, 1981). The masculinist construct of the discourse excluded women from full participation in the elite. Thus, whilst there are numerous references to women's charity work and their role as social hostesses in contemporary accounts, even aristocratic women do not have a political presence of their own. Not one woman is included among the nearly 150 local leaders listed in the four Somerset volumes.

Secondly, the discourse of the country gentlemen promoted the notion of *stewardship*. Deprived of the military duties of the medieval knight, the gentlemanly ethic had reinvented the role of the gentleman to be the stewardship of the countryside. This was a responsibility that could only be discharged through the ownership of land, and, significantly, embraced not only the stewardship of the physical environment, but also the stewardship of rural society and rural people (Everett, 1994). As such, the leadership activities of the country gentleman were recast not as a privilege, but as a duty.

Thirdly, the discourse of the country gentleman reproduces the idea of a 'natural order' within society. This extends the predatory order recognized in nature such that humans are represented to have supremacy over the natural world and 'nature's most ancient peerage' (Press, 1890, p. 22), are given supremacy over other rural dwellers. Political leadership is thus tied to 'nobility' and hence to the hereditary principle. Positions of power are not offices to be competed for, but rather are bestowed by birth-right.

Finally, the discourse promoted a particular moral geography in which the nobility of the country gentry was portrayed as a reflection of the nobility of the countryside. Resistance to urban influence, as part of a defence of rural purity, was paralleled by the protection of the purity of the rural elite by the virtual exclusion of individuals whose wealth derived from commerce. The few who were accepted, such as the Chocolate-manufacturer, Francis Fry, obscured their background so as not to conflict with the dominant discourse and only one of the 114 entries in Gaskell (1906) mentions any commercial activities – that of Francis Meade, a county councillor, magistrate, coroner and Portreeve of Langport who was a wholesale grocer and cheese manufacturer.

In contrast, no members of Somerset's most successful mercantile families – the Clarks, Foxes and Wills – were included in the volume, despite both William Clark and Charles Fox serving as aldermen and magistrates and two members of the Clark family sitting as county councillors. The Wills family, furthermore, had by 1906 acquired many of the trappings of the country gentry. They were significant landowners; three members of the family had been made baronets, two had sat in parliament, and Sir William Wills had been High Sheriff of Somerset in 1905. Yet, it was not until the publication of the fourth of the volumes in 1908 that Sir William was finally included, after his ennoblement as Lord Winterstoke.

The explicit presentation of the elite in the directories of 'Somersetshire Leaders' was replicated in the landscapes of power that they created. Large country houses and their landscaped parks, such as Fairfield House at Stogursey, Poundisford Park, and the Court House at East Quantoxhead, would have been very visible expressions of power in a less urbanized, less mobile, society. The most extreme example of visible power was at Dunster, where the Castle – home to the Luttrell family – literally dominated both the village and the surrounding countryside.

Hunting and Associational Power

The discourse of the country gentleman was also performed through the elite's favourite past-time, hunting. Within the cultural norms of the time, hunting was accepted both as a 'natural' activity, which demonstrated the country gentleman's closeness to nature, and as a unifying force supported by all parts of society (Newby, 1987; Woods, 1997). However, working class participation in the hunt was strictly regulated by an internal stratification that reflected the class structure of wider rural society. To participate as a full member of the hunt required both economic and cultural capital and was hence restricted to the gentry and their equivalents. The farm labourers and estate workers who joined the hunt did so as habourers, whips and huntsmen – the footsoldiers.

Hunting was, therefore, patently an elite sport that in its most prestigious manifestations formed a rural equivalent of the London 'season' (Newby, 1987). During the latter part of the 19th century, hunting guests at Dunster Castle included the Lord Chancellor, the former Prime Minister, William Gladstone, and, in 1879, the Prince of Wales (Woods, 1997). More routinely, the hunt formed a meeting place for a large number of the county's social and political elite:

> Sir William Karslake has for many a year been a member of the Hunt Committee, varying the toil of government at Somerset House with the welcome relaxation of a gallop over the heather and the grass; Lord Poltimore is to be seen at the Cuzzicombe Post meets, which lie nearest to his North Molton residence at Court Hall; Sir C. T. D. Acland, the owner of the field and of some very large slices of red deer land, is generally to be found at the first meets, and Mr. Luttrell has not far to come from Dunster Castle; whilst Viscount Ebrington, the owner of Exmoor proper, chairman of the Hunt Committee ... is sure to be present ... the member for West Somerset, Vice-Chamberlain and Treasury Whip, Sir Alexander Acland Hood, had traveled from Saint Audries; Mr. Basset, of Watermouth Castle, and former Master of the Staghounds, is to be seen; and the Baroness Le Clement de Taintegnies is dispensing hospitality to an admiring circle ... The master, Mr R. A. Sanders ... contested the Eastern division of Bristol at the General Election of 1900, and considerably lowered the previous Liberal majority. In 1901 he became an alderman of the Somerset County Council. (Evered, 1902, pp. 31-32).

Indeed, Sir Robert Sanders credited the contacts he formed as Master of the Devon and Somerset Staghounds with assisting the subsequent development of his political career, in which he became MP for Bridgwater (1910-23), and later for Wells (1924-29), Minister for Agriculture, a Privy Councillor, and, after being raised to the peerage as Lord Bayford, Chairman of Somerset County Council (1937-40) (Bayford, 1984).

Hunting was just one forum through which the associational power of the Somerset elite was maintained. Many of the social and political leaders of the county were related to each other, with dynasties such as the Acland-Hoods, Frys, Portmans and Luttrells/Somervilles represented across a range of local government positions. A number more had been together at boarding school or university: Henry Hobhouse and Sir Alexander Acland-Hood had been contemporaries at Eton and Balliol College before sitting on opposing parliamentary benches for neighbouring constituencies; whilst Arthur Fownes Somerville, a magistrate and county councillor, had been at Eton with the future Judge Copplestone and at Trinity College, Cambridge, with the Earl Waldegrave. The county's aristocracy and senior gentry, meanwhile, were almost all members of London gentlemen's clubs, notably Brook's, Arthur's, Bachelor's and the Carlton Club (Gaskell, 1906). As such, whilst elite members stressed their local identity in building discursive power, its associational power significantly depended on interactions in spaces located outside the county.

Civic Power and Agrarian Leadership

The end of the squirearchy's dominance came not in any sudden revolution, but in a series of events that gradually chipped away at the three dimensions of the landed elite's power. The first major challenge came from the Liberal government of 1906-14. Although the government's benches included many members of the rural landowning classes, and although it failed to go as far as implementing radical policies for land nationalization or land taxation, it did introduce more modest land reforms that made it easier for tenants to purchase property, thus eroding a little of the elite's resource power. Moreover, the constitutional show-down of 1911 between the government and the House of Lords scratched some of the gloss from the discourse of the country gentleman. If the elite survived these challenges merely shaken, the next event was to prove far more serious.

The aristocracy and landed gentry had always exhibited a strong military tradition and the vulnerability of the junior officers who swapped the estates of rural England for the trenches of Flanders meant that the rural elite suffered a disproportionately heavy loss of life in the First World War. Not only did this threaten the hereditary principle, but families were also faced with demanding death duties. As the value of agricultural land increased sharply in the first few years after the war, the sale of land became the most attractive way of meeting these debts.

In fact the dismantling of the great estates had begun before the war, prompted in part by land reform and particularly the continuing threat of land taxation. Between 1910 and 1915 around 800,000 acres, worth approximately £20 million, were sold. It was in 1919, however, that the flow of land sales became a torrent. Howkins (1991) notes that over half a million acres of land had been put on the market by March 1919, and in total around a quarter of the land surface of England is estimated to have changed hands between 1918 and 1922 (Beard, 1989).

Some landowners simply trimmed their land holdings, selling off parts of their estates to their tenants but compromising their resource power in doing so. Other aristocratic families that had maintained estates in a number of different counties embarked on a more radical rationalization, effectively 'deterritorializing' from entire counties as whole estates were sold. In Somerset the Portman family, which in 1871 had been the county's largest landowner with over 24,000 acres, sold up and withdrew from the political life of the county. A few landowners went even further, leaving Britain for the colonies, as Lee observed in Cheshire:

> Some of the landed families were prepared to leave England in the hope of continuing to enjoy the pleasures of a country estate. In the early 1900s Lord Delamere of Vale Royal was largely responsible for starting the first white settlements in Kenya, Lord Egreton of Tatton, and in the 1930s, Sir Henry Delves

Broughton, followed his example. Here the county lost two of the old 'country party' families. (Lee, 1963, p. 92).

The sale of land did not only undermine the resource power of the squirearchy, it also eroded their associational power and discursive power. There were now less landed families around, and many members of those who remained could no longer afford to live lives of leisure mixed with public service, but were compelled to seek paid employment. The social calendar of the rural elite was hence hit by the diminution of participants. Lee (1963) records that the last dance of the Knutsford County Assembly, an institution of elite life in Edwardian Cheshire, was held at Arley Hall in December 1930. Similarly, Claude Luttrell, the fourth son of George Fownes Luttrell of Dunster Castle, who had been 'forced' to find employment as a director of the Westminster Bank, wrote a book for others in his predicament, 'destined to earn their living at some uncongenial work' (Luttrell, 1925, p. 11), who found that work got in the way of hunting and shooting (Woods, 1997). In these ways many of the spaces and events through which the associational power of the elite had been reproduced contracted or simply disappeared altogether (Beard, 1989).

Perhaps most significantly, as land was sold, links with communities broken, social activities sacrificed and the vulnerability of the elite exposed, the discourse of the country gentleman could no longer be sustained. Rural Britain was in need of a new political settlement.

That is not to say, however, that the aristocracy and gentry disappeared from local power structures in rural areas such as Somerset. Of the 25 alderman of Somerset County Council in 1935, ten were landowners, including three members of the Hobhouse family and two peers (Lord Bayford and Lord Strachie, both former MPs). Of the 74 county councillors, 17 were landowners. Members of the aristocracy and gentry played a prominent role in the establishment of Rural Community Councils during the inter-war period, including the Marquess of Bath and Major Cely-Trevilian in Somerset (Woods, 1997), and continued to be well represented on the magistrates' benches and to dominate the ancient offices of Lord Lieutenant, Deputy Lieutenant and High Sheriff. Moreover, a culture of deference persisted such that those members of the landed gentry who did continue to be involved in local government were looked to as natural leaders, as descriptions of two senior Somerset councillors of the 1970s indicate:

> The then leader of the Conservative Group, Penny Phillips, a former, I think she must have been something in the Resistance during the war, she had the Croix de Guerre or something of that nature, a classic lady of spirit you know, who sort of runs the family estate, she owns most of the Blackdown Hills ... But very much the classic case, the landed gentry but concerned for the well being, of, you know, her tenants, the population. She'd been given by, you know, fate, time and resources to devote herself to public service. Now, whatever her policies, that attitude I can cope with and I can understand, you can live with that. (Interview I255 – Borough councillor).

There was a splendid chap here called George Wyndham, one of the Wyndham estates, a very important landed family. Now, George Wyndham was a first at Oxford, he was then in and worked with the Indian Civil Service for virtually a lifetime, you know, half a lifetime, put it that way, and he had considerable territorial interests in Somerset, his roots were very much down in the county. Now a chap like that would feel that he had nothing to gain from it, he would just feel that with his experience that perhaps he could be of use. And he became chairman of the county council in his turn. (Interview 1190 – Former county councillor).

What the old elite could no longer do, however, was to rule alone. It needed to rebuild its associational power and did so by opening up its elite networks to engage with (though not incorporate) members of two strengthening political groups – farmers and the small town business class.

The Rise of the Agrarian Elite

As the old landed elite retreated the first beneficiaries in both economic and political terms were the farmers who purchased their tenancies in the great sale of land. Thompson (1963) described the transfer of property from aristocratic landlords to small-scale farmers as 'a startling social revolution in the countryside, nothing less than the dissolution of a large part of the great estate system and the formation of a new race of yeoman' (p. 333), and Howkins (1991) suggests that whilst the farmer-landowners were minor figures in national terms, locally they were 'masters of the parishes where they lived, perhaps more surely than those who had owned a parish only as one of many hundreds. The focus of power and deference finally shifted in these years from the landowners to the farmers' (p. 281).

The local basis of the farmers' power was demonstrated by the fact that their colonization of rural local government was most pronounced at the most local level. Farmers came to dominate the membership of parish councils and it was not uncommon for one of the large farmers in a community to chair the parish council for several decades, in many cases handing over to their son on retirement. One study of 21 parishes in South Devon found that three in ten parish councillors were farmers (Mitchell, 1951). Farmers were also commonly elected as the parish representative on the rural district council. Bracey (1959) suggests that farmers constituted over 50 per cent of members of rural district councils in the 1950s, whilst the Royal Commission on Local Government reported that 35 per cent of rural district councillors in 1967 were farmers. The influence of farmers on these bodies was further strengthened by the relative absence of the gentry, although Bracey (1959) remarked that, 'a rural district council may consider itself fortunate if it has one councillor from the upper class who by his upbringing and experience can assure a reasonably high standard of conduct at meetings' (p. 75).

The presence of farmer councillors was less marked at the county level, where they had to compete for influence with small town business elites, the rump of the old landed elite and, in more industrialized counties at least, trade unionists. Nonetheless, the representation of farmers on county councils increased steadily during the first part of the 20th century. In Somerset, for instance, farmers comprised 12 per cent of the county council in 1906, but 15 per cent in 1912 and 20 per cent in 1930.

The rise of this new agrarian elite in part reflected the growth of farmers' resource power as both employers and, increasingly, owners of land. Economically, too, farmers had done comparatively well from the wartime period, increasing their relative wealth within rural communities. Guaranteed prices during the First World War created financial stability and farmers were able to significantly increase their net income (Howkins, 1991). Agricultural prosperity, however, came to a halt in the 1920s with a sudden depression, thus providing farmers with a reason for political mobilization. Farmers also benefited from a transfer of discursive power. Although the discourse of the country gentleman had been discredited, many rural residents still held to the underlying rationale that leadership should be exercised by the significant owners of land and therefore as the gentry sold up they looked to the larger farmers as their new leaders.

The main obstacle faced by farmers was that they did not have an equivalent to the social networks of the landed elite through which associational power could be maintained. This gap was filled, however, by the organizational role assumed by the National Farmers Union (NFU). Founded in Lincolnshire in 1906 as a body to defend the interests of tenant farmers against the twin threats of farm labourer trade unionism and landowner pressure, the NFU came of age during the First World War when the need to secure the food supply took the union into policy-making circles (see also chapter six). By 1918, membership of the NFU had reached 80,000. After the war, the NFU developed its influence at all levels of rural government, particularly through a close relationship with the Conservative party (see chapter four), but also including an explicit strategy of increasing farmer representation in local government (Hallam, 1971). As Lee again records in Cheshire, NFU office holders became key players in both local government and the structure of local agricultural administration:

> The Cheshire branch of the National Farmers Union developed its own tactics as a pressure group. Its ideas found expression in the formation of an Independent Party on the County Council at the end of the First World War ... The Independent Party was organized by J. O. Garner, the local secretary of the NFU ... T.C. Goodwin, chairman of the County Milk Committee, who was elected county councillor for the Church Coppenhall division in 1919, and John Done, elected for the Malpas division in 1925, were important leaders of the local branch of the NFU ... Thomas Peacock, who became chairman of the Milk Marketing Board, was atypical local leader thrown up by this movement, the pressure group politics of farmers. He was elected a county councillor for the Tattenhall division in 1931, and was an unsuccessful National Liberal candidate for Parliament in the Eddisbury by-election of 1943. (Lee, 1963, pp. 96-7).

Through these tactics the NFU succeeded in helping farmers to assume a central position in the local power structures of rural Britain. This new found influence was symbolically recognized in the change of the NFU's motto from the original 'Defence not Defiance' – a slogan that Hallam (1971) suggests 'conjures up the picture of a tenant farmer, rather in awe of the gentry and in particular of his landlord' (p. 174) – to the more imposing Latin phrase, 'Labore agricolae floreat civitas', or 'May the State prosper through the farmer's toil'.

However, farmers were not the only members of rural communities to step into the leadership vacuum left by the retreat of the gentry. Positions of leadership were also occupied by prominent figures in what Grant (1977) calls 'political occupations'. These are positions that involve an individual in regular, high-visibility interaction with a wide range of local residents. Thus, Grant identifies vicars, veterinary surgeons and auctioneers as 'political occupations', but to this list might also be added shopkeepers, postmasters and mistresses, doctors, solicitors, teachers and publicans. The leadership status of such persons is hence primarily a product of their individual social capital and their ability to mobilize sufficient associational power to secure election in a contested ballot. Grant, for example, argues that persons in political occupations, 'may have a sufficiently wide range of personal contacts to garner large number of votes from electors who know them personally' (1977, p. 17), whilst a study of Kincardineshire in the 1960s suggested that 'elections are something of a personality contest in which the candidate with the most friends and neighbours wins' (Dyer, quoted in Grant, 1977, p. 17).

Leaders from political occupations also benefited from the mutually-supporting combination of resource power and discursive power. They were credited as being suitable leaders because they were considered to be easily accessible, well known to local people, knowledgeable about local affairs and clearly identified with the community. These are 'resources' which were discursively constructed as desirable for a local leader to possess.

The Civic Business Elite

The leadership role performed by persons in political occupations in rural communities was paralleled in the domination of local government in market and county towns by the 'shopocracy' (Elcock, 1975) of traders and businesspeople. Commerical interests had long exercised influence in civic government, but their significance was increased in the inter-war period by the growing activity of municipal councils, and by the opportunities that arise to play a greater role at a county level as the influence of the gentry waned. Councillors from commercial and professional backgrounds continued to comprise around a quarter to a third of county councils such as Somerset, but began to exert more influence on the council even if senior positions still resided with the landed members. Within the towns, however, the dominance of the business community approached a virtual monopoly. At least 16 of the 24 members of Taunton Borough Council in 1928 were businessmen trading in the town, including three builders, two shopkeepers, and one tailor, solicitor, corn merchant, dairyman, bookbinder, fishmonger, builders' merchant, timber merchant, photographer, coach builder and insurance broker apiece.

The position of the small town business elite once again built on the three dimensions of resource power, discursive power and associational power. In a similar way to the rural political occupations, traders and business-owners were credited with being accessible, visible and having a wide range of contacts. Successful businessmen also had wealth that enabled them to participate in 'public service', as well as control over their own time. Moreover, they were empowered by the reproduction of a discourse that conflated the interests of a town with the interests of business.

Such factors marked out business owners as prospective local leaders, but the extent of their dominance of civic governance indicated their success in mobilizing associational power through a close-knit, exclusive elite network. The over-lapping social networks of civic leaders were charted in 1940s Banbury by Margaret Stacey who recorded inter-locking clusters of membership by Conservative and Liberal borough councillors of churches and chapels and sports and social clubs (Stacey, 1960). More explicitly, organizations such as the Chamber of Trade, Rotary Club and Freemasons had memberships that strongly correlated with that of the business elite and formed an 'elite space' through which associational power could be maintained and exercised (see also Woods, 1998b). Taunton Borough Council in 1928, for example, included among its members the secretaries of two of the town's Masonic lodges.

Some members of the small town business elite also participated in wider social networks that connected them to the agrarian and landed elites, hence contributing to the production of the associational power that underpinned the new rural political settlement. In Somerset, foxhunting formed one such meeting-point, as demonstrated in 1934 when the Taunton Vale Foxhounds held a meet in the centre of Taunton in honour of Mayor William Brake, a keen hunt member (Bush, 1988). Businessmen who traded with farmers and landowners also provided pivotal links, exemplified in Somerset by two Robert Brufords, father and son. A farmer and businessman, the elder Robert Bruford was a long-serving member of Taunton Town Council and of the Board of Guardians, despite accusations of public drunkenness. His son, also a farmer and businessman, became in sequence, a rural district councillor (and later chairman), a county councillor for a rural ward, chairman of the Taunton Board of Guardians and a county alderman. In 1922, he was one of the first farmers to be elected to parliament, as MP for Wells, his selection as a candidate having been remarked on as a radical departure in the local press (Woods, 1997). Bruford junior served as a member of the Somerset War Agricultural Committee and through these offices was a key figure in the emergent agrarian elite in the county, but he was also an active Taunton businessman as chairman of the Taunton Gas Company, of the West Somerset Dairy and Bacon Company, and of the Priory Land Development Company.

The Organic Community and Conservativism

The inter-war political settlement that placed rural power in the loosely affiliated landed, agrarian and small town business elites was made possible not just by the kind of personal connections described above, but also by the subscription of all three groups to a broadly conservative political ideology. All three groups had a vested interest in the maintenance of property rights and of capitalism, and in resistance to any broadening of political participation that would disturb the entrenched class system. As such, all feared the growth of trade unions and of the Labour party in rural areas. At the same time, all legitimized their own leadership as a performance of 'public service' that the middle classes had a duty to undertake. As one veteran councillor in Somerset recalled:

> I was a day boy at a public school and I left I 1939, and on the day I left one of the masters ... he said, 'When you go out you will always be a leader. People will look to you because you've had this education', he said, 'although you are now fitted for ordinary [life] ... no doubt making yourself lots of money etcetera', he said, 'don't forget your real... one of the other great purposes in life is service to the community.' And that has influenced me a great deal. When I came out of the war, the navy, I decided then that I would give half my life to making money, here in this business, and half to public service. (Interview – I50, Borough Councillor).

The elites were also influenced by new discourses of rurality. They drew on the ideas of the 'back to the land' movement and the Ruralists (Howkins, 1991; Matless, 1990, 1994), which called for the 'preservation' of rural England in an anti-urban celebration of 'tradition' and 'rural community'. Yet, agrarian Conservatism also reacted to the economic pressures on agriculture in the 1920s and 1930s by pushing agriculture to the heart of the construction of rurality, a centrality that included policy commitments to infrastructural improvements, both locally and nationally. This commitment was in turn matched by the small town business elites who enthused about the idea of 'progress', supporting projects that improved public utilities and infrastructure, built new public buildings and modernized the urban landscape.

These threads came together in the discourse of the organic, agricultural, community. Unlike the earlier discourse of the country gentleman, the discourse of the organic, agricultural, community was primarily a discourse of rurality rather than a discourse of power. However, it informed discursive power in three ways. Firstly, the centrality given to agriculture strengthened the claim of farmers and landowners to political leadership, whilst the positioning of market towns as service centres for their rural hinterlands strengthened the 'shopocracy'. Secondly, the motif of 'community' encouraged both the cohesion of former estate villages and the identification of large farmers and people in political occupations as the new leaders to replaced the vanished squire. Thirdly, by presenting the rural community as a place of stability, tranquility and solidarity, the discourse exaggerated the urban/rural contrast and promoted the notion of the 'apolitical countryside' as discussed in chapter one. The discourse of the organic, agricultural, community therefore painted a gloss of stability over what was in effect an environment of social, economic and political change. Indeed, the notion of 'stability' was being promoted by organizations that were significant agents of restructuring: the NFU, the Conservative party, and, increasingly, the county council.

Pageantry and Politics

The themes of community and tradition embodied in the discourses promoted by the civic and agrarian inter-war elites, and their effect in detracting from a process of considerable social and economic restructuring, were all evident in a series of local pageants that were held in towns and cities across Britain, including Taunton, in the first half of the century (Woods, 1999). Following a format devised by Louis Napoleon Parker, director of the first pageant in Sherborne, Dorset, in 1905, the pageants combined music and drama in recounting local history through a number of acts focused on different historical events. The stylized presentation of these events reflected Parker's intention that the pageants act as a form of historical morality play or 'folk-play'.

The Taunton pageant was staged for six nights in the summer of 1928. It involved 1,500 local performers and was attended by audiences of 3,000 people each night. Entitled 'Defendamus' – 'let us defend' – the pageant presented a narrative of Taunton's history that emphasized a single underlying theme, of 'continuity', as the Pageant Master, Major Maurice Cely-Trevilian – a local landowner, magistrate and High Sheriff of Somerset in 1925 – explained in a talk to the Taunton Rotary Club:

> One idea runs through it – that of continuity – that we are not standing in our own generation, but that we are part of the productivity and aspiration and great deeds of those who have gone before, and that we have to pass them on to our children afterwards. (Major Cely-Trevilian, quoted in the *Taunton Courier* 6/6/28).

This sentiment is reminiscent of the beliefs embodied in the ideal of the 'country gentleman'. Continuity was political as well as social and moral; and all these ideas were encapsulated in the closing stanza of the performance:

> And Taunton's children through the centuries,
> Have learnt the meaning of Defence
> So as ye pass from here towards thy homes,
> Which prove a thousand years of work and love,
> Make high resolve that you will show yourselves,
> Worthy of those who passed this way before,
> Firm in Faith, strong to defend the Right.
> The past is past; the future dawns. Good-night.
> (Nimue, Epilogue).

A second theme running through the pageant was that of localism. The whole structure of the performance was designed to emphasize Taunton's importance and inspire pride in the town's history. The script alluded the favourable geography of the town, the character of King Ina complementing the setting in founding the town. Furthermore, the accompanying publicity to the pageant sought to position the very performance and success of the pageant as being inseparable from the identity and reputation of town itself. Thus, as Cely-Trevilian told the Rotary Club:

> Let it be remembered that the Pageant is a Taunton venture and that the reputation of Taunton is at stake with regard to the entirely adequate produce of the show. (Major Cely-Trevilian, quoted in the *Taunton Courier*, 6/6/28).

The strategy of localism had a dual purpose. Demonstrating civic pride and civic achievement was vital to Taunton's attempt to market itself as the 'county town' of Somerset, having only recently won the competition to be the location of the new county hall, which was built on the site of the pageant, opening in 1935. Equally, the stress on localism aimed to re-create a sense of community that supported traditional social formations in contrast to the class politics espoused by socialism. Both these concerns were evident in the *Taunton Courier's* coverage of the pageant:

> Taunton, the county town of Somerset, is this week a very gay and proud borough. Not only are 1,500 inhabitants of all ages and walks of life engaged nightly in staging a grand pageant, depicting twelve centuries of Taunton history, but residents generally are also showing a very live interest in the event. (*Taunton Courier*, 27/6/28).

However, despite the superficial rhetoric of cohesion, the notion of community reproduced by the pageant was one ordered by an elite power structure. The script of the pageant speaks of all classes assembled to 'do homage' to the mediaeval lord of the manor, a culture of deference that is also explicit in the frontispiece to the pageant programme which shows the 'presentation of bible sword and colours to the Duke of Monmouth by the fair maids of Taunton Deane'. The illustration also emphasizes the gendering of the pageant, showing female subservience to masculine power. Female parts are restricted either to the wives of male characters, charged with injecting comedy into the script, or to mythical characters such as the Lady of the Lake, the Queen of the Wasteland, the Queen of Northgalles, Morgan Le Fey and Nimue. These mythical characters represent both a deeper wisdom and a containable feminine power, distanced from the historical narrative – where the victories are won by heroic males - by both their dramatic restriction to the prologue, epilogue and scene changes, and their dehumanized form. This representation again resonate with a popular discourse which equated femininity with Nature – and by extension linked the human realm with male leadership, a provocative discourse for the Taunton Pageant to reproduce at a time when the press was hotly debating the extension of the female franchise.

Similarly there was only one woman – the president of the choral society who was also the wife of a local landowner – amongst the 25 patrons and vice-patrons, and 41 members of the two main organizing committees. Whilst the names of the actors were not printed in the programme, it did list all of those involved at the organizational level – all of whom were drawn from the ranks of the elite. The patron was the brother of the King, the Duke of York. The vice-patrons included the Lord Lieutenant and the High Sheriff of the county, local landowners, the local Bishops, the Member of Parliament and former MP for Taunton, the chair of the County Council, the commanders of local military units and the mayors of neighbouring towns. The town council formed the executive committee, chaired by the mayor, Howard Westlake, whose portrait was featured in the programme, and the general purposes committee was composed largely of prominent local businessmen.

The elite structuring of the pageant was further emphasized by the opening ceremony which included a 'civic procession' from the municipal buildings to the pageant site, led by the mayor of Taunton accompanied by the mayors of neighbouring towns; and by the pageant festival service held on the preceding Sunday at the town centre church of St Mary. Lessons were read by Cely-Trevilian and Mayor Westlake, and the vicar exhorted his congregation to 'Help us build Jerusalem's walls here in this pleasant vale in which we live', alluding to the English patriotic hymn, Jerusalem, which was sung at the close of the pageant, and whose lyrics – 'And did those feet in ancient time/ walk upon England's mountains green?' – both resonate with the invented nineteenth century mythology that Christ visited Somerset as a child, and were rich in symbolism for the localist and nationalist discourses of the elite.

Moreover, whilst the pageant embraced transgressive elements of the carnivalesque, allowing tradespeople to act as 'king' or 'queen' for the short period of the performances, the script firmly reinforced the importance of the hierarchical social order. This was most explicitly conveyed through the penultimate stanza of the performance that reminded participants that the transgression was temporary and called on them to return to their stations in life:

> And so our Pageant's done; back to the parts,
> Which now they play on stages past your ken,
> Press king and bishop, soldier, serf and reeve –
> Leaving their mark upon the ways they trod –
> Leaving their tools all ready to your hands
> To raise the age-old walls another course.
> (Nimue, Epilogue)

As such, rather than acting as a catalyst for a new, less class-based society, the pageant created an illusion of democratic, class-less involvement, whilst at the same time propagating discourses that supported the existing social order. The Taunton pageant, like those held in other towns, can be regarded as an event which was not only organized and sponsored by the local elites, but which was also explicitly and deliberately employed as a mechanism for reproducing ideas of continuity and community that resonated with the elites' discourses of power and locality.

The Post-War Years

The structure of local power in the British countryside that was constructed during the inter-war years, with farmers dominant in rural communities and united in a broad coalition with the remaining landed gentry and the small town business elite at a county scale, persisted in recognizable form well into the latter part of the century. The years after the Second World War did, however, begin to see a number of processes of change that would eventually end the agrarian-business hegemony.

Firstly, the scope and responsibilities of local government expanded significantly, in part due to the establishment of the 'welfare state'. As the management of local government became more complex, so more power and influence passed from the councillors themselves to the growing professional cadre of local government officers. Indeed, the relative influence of council officers (and of a small minority of more politicized councillors, particularly from the Labour party), was exaggerated by the continuing spirit of amateurism among the dominant elite and the indiscipline of councillors. As one former Somerset county councillor recalled:

> Those who had planned what their moves were going to be were able to drive a coach and horses through ... because the old village squires might be people who if they heard a good argument or a well presented argument they could be swayed by it, and so you got all these splendid squires bumbling up in their garters and wandering into the council chamber and voting all over the place. (Interview – I156 Former county councillor).

Secondly, the late 1940s and 1950s saw the emergence of a new leadership group within rural society comprised of ex-military officers. Senior officers, many of whom had retired comparatively young, brought considerable leadership experience and expertise to the councils and other bodies which they joined as their 'public service' in retirement. Members of Somerset County Council in the post-war period, for example, included Marshal of the RAF Sir John Slessor, Lt. General Sir Reginald Savory, Brigadier Eric Frith, Admiral Sir Hugh Tweedie, Brigadier R. Austin, whilst from a later generation of military leaders, Air Vice Marshall Harold Leonard-Williams was chairman of the council in the early 1980s Many senior officers came from landed backgrounds, whilst others were, as the new occupants of large rural properties and people used to deference in a hierarchical structure, natural allies of the traditional elite into which they were absorbed. At a junior level, Bracey (1959) identified the non-commissioned officer 'who, having gained experience of leadership in the armed Services, has returned to take over the village stores or pub or to travel into town each day to work' (p. 52) as a potential village leader. As he commented, reflecting the discourse of the time:

> Some of these people occupy positions of authority by virtue of their natural qualities of leadership. They do not seek leadership but have leadership thrust upon them. Any satisfaction they derive from the positions their special abilities have won for them is incidental. (Bracey, 1959, p. 52).

Thirdly, agricultural modernization from the 1948 Agriculture Act onwards gradually undermined key elements of the farmers' resource power. Most significantly, mechanization and the introduction of biotechnologies reduced the agricultural labour force, such that the position of farmers as major employers in small rural communities was diminished. Other changes included the amalgamation of farm units, reducing the number of farmers, and the growth in corporate agriculture, with farms run by managers who were less disposed than own-account farmers towards community leadership. Furthermore, the increasing production of agricultural goods under contract to food processing companies or supermarkets, and associated decline in local sales, contributed to a weakening of the ties between farm and village.

Finally, the trend of counterurbanization from the late 1960s onwards provided the perhaps the greatest challenge to the rural political order by bringing into the countryside large numbers of in-migrants who had not been raised with the discursive conventions of rural society and had few connections to agriculture or country sports, but who often possessed skills and interests that led them to seek a role in local leadership. Bracey (1959) noted the arrival of 'elderly middle-class people who have retired to the country', some of whom were 'prepared to help things along' as early as the 1950s, observing that:

> In some villages they slip easily into leadership of different local organisations thereby brightening their own lives at the same time. Elsewhere, the newcomers are kept in their place, namely on the fringe of the village community, by the still all too-frequent remarks which carry so much weight in village conclaves, such as 'We've lived here for fifty years and we do know it won't work', or 'It's always been good enough for us', with great emphasis on the 'us'. (Bracey, 1959, pp. 51-2).

Generally speaking, this first wave of in-migrants were drawn from the more affluent reaches of the middle classes and shared the broad conservative outlook of the rural elites to whom they presented little threat and even, on occasion, were able to join. As urban to rural migration intensified in the late 1970s and 1980s, however, and as the social status of in-migrants became much more inclusive, so the potential for conflict with the existing agrarian elite increased. It was these last two trends – agricultural modernization and counterurbanization – that formed the background to the classic study of local power in East Anglia by Howard Newby and colleagues that is discussed in the section below.

Property, Paternalism and Power in East Anglia

The agricultural geography of East Anglia has long influenced its local politics. The predominance of arable cultivation, which was in turn associated with large farm units and significant workforces, meant that the large farmer of Norfolk or Suffolk occupied a position of power within their community that was more akin to that of the squire than that of the small farmer in other parts of the country. At the same time, however, the large number of agricultural workers in the region created a potential for trade unionization, and indeed helped to support the election of several Labour MPs for rural East Anglian constituencies in the 1940s and 1950s. Overall, though, the political scene examined by Newby *et al.* in the early 1970s was one in which farmers continued to be the most prominent group. In 1973, farmers comprised 21 per cent of councillors at all levels in East Anglia, and 31 per cent of rural district councillors. A further four per cent of councillors were part-time farmers and five per cent were directly involved in the agricultural industry in some other way. Following the re-organization of local government one year later in 1974, farmers still comprised 16 per cent of the members of Suffolk County Council, 11 per cent of Suffolk Coastal District Council and 35 per cent of Mid Suffolk District Council (Newby *et al.*, 1978). Moreover, farmers dominated the key positions on these councils. On Suffolk County Council, for example, farmers,

> enjoyed something approaching a monopoly over key positions. Thus the chairman and vice-chairman of the county council were both farmers. So too was the leader of the majority Conservative group (who was also leader of the council). The chairman of the Planning Committee owned and farmed over a thousand acres in Suffolk, as did the Education Committee chairman. The chairmanships of the key finance and policy sub-committees were both held by farmers, while the chairman of the Social Services Committee was a large landowner. There were, in fact, no major committees on the county council which were not chaired by farmers or landowners. (Newby *et al.*, 1978, pp. 232-3).

Neither was it only in elected local government that farmers exerted their influence. Over two-thirds of the farmers owning more than a thousand acres surveyed by Newby *et al.* held positions of responsibility in the local community including service as school governors, magistrates, trustees of local charities, and as chairs of village hall committees and parochial church councils. This hegemony marked the enduring legacy of the rise of agrarian power in the years after the First World War. Most of the farmers may not have come from the stock of the landed gentry, but many played the role of village squire, either through choice or expectation:

> I can name several ... They are JPs, pillars of society and the church, school governors. They're feudal. Their approach to labour is that labour should be grateful to have a job for which they should grovel in return. They think they have a divine right to rule. (Norfolk farmer quoted by Newby *et al.*, 1978, p. 197).

> Farmers are forced to behave like squires because people still expect it. In a village like this, people expect you to lead. (Suffolk farmer quoted by Newby *et al.*, 1978, p. 197).

Indeed, Newby *et al.* found that the farmers who were active in local leadership were not representative of farmers as a whole, but tended to be the larger landowners and employers. Nearly three-quarters of the farmer-councillors in 1973 employed labour, compared to just under a third of all East Anglian farmers; 42 per cent employed five or more workers, compared with 8 per cent of all farmers in the region. As such, Newby *et al.* (1978) note, 'we are looking at much the same class of farmers as that which has dominated rural politics for centuries' (p. 231).

Although Newby *et al.* do not explicitly use the language of resource power, discursive power and associational power, all three dimensions can be identified in their analysis. Adopting a political economy stance, Newby *et al.* emphasized relations of class, property and labour in explaining the rural power structure, arguing that farmer participation was motivated by a desire to maintain favourable conditions for agricultural production and a supply of farm labour. The ownership of property, they argued, was the key determinant in the class system of rural Britain, and hence in its power structure. This prioritizing of property was in turn supported by particular 'ideologies of property ownership' which 'contribute to a system of "natural" inequality in the countryside which can remain an extraordinary prevalent feature of the taken-for-granted perception of rural society' (Newby *et al.*, 1978, p. 25, see also Rose *et al.*, 1976).

Property requires maintenance, and therefore many landowners were also employers of labour. However, as Newby *et al.* describe, the agricultural economy of East Anglia produced a particular type of labour relations that were characterized by particularistic, face-to-face, interaction between the farmer/employer and the worker, in turn supporting a 'complex web of traditional authority relationships' (p. 26), that they label 'paternalism'. For Newby *et al.*, paternalism 'enabled power relations to become moral ones' (p. 28) by representing the landowner or farmer as the benevolent steward of rural society who was owed deference by their labourers who were often totally dependent on them for both work and housing. It thus involved both 'autocracy and obligation, cruelty and kindness, oppression and benevolence, exploitation and protection' (p. 29). Yet, Newby *et al.* contended, paternalism was not just about cultural values, it was grounded in and dependent on the concrete social relations of rural society:

> Paternalism does not exist in a social vacuum – it is derived from and embedded in a particular system of social stratification, the source of which is basically economic and objectified through property. Paternalism is therefore a method by which class relationships become defined, and grows out of the necessity to stabilize and hence morally justify a fundamentally inegalitarian system. Paternalism – and its obverse, deference – must therefore be regarded as a relationship rather than an attribute of the parties involved. (Newby *et al.*, 1978, p. 29).

The particularistic nature of labour relations also helped to secure the agrarian elite's hold over associational power by militating against the organization of working-class interests. In a system where most councillors sat as independents and the role of political parties was weak, the emphasis was placed on direct, individual, face-to-face interaction with councillors. As Newby *et al.* noted, the quality of access that working class residents could obtain to councillors was considerably inferior to that of the middle class residents who shared the same social networks as leading figures in local politics. Overall, the effect was that attention was directed away from general grievances towards individual complaints and thus collective interests of the rural working class were 'constantly undermined through the operation of a political system where only individual interests are recognized' (Newby *et al.*, 1978, p. 269).

The product of the local power structured observed by Newby *et al.* was the perpetuation of a fundamentally unequal rural society. Through the pursuit of policies that resisted industrial development, restricted council house building and prioritized the setting of low rates, the farmer-dominated political elite in East Anglia were successful in preserving the countryside as an agricultural space, thus upholding their discursive power, and serving the economic interests of agrarian capitalism. Resistance to industrial development limited the opportunities for alternative employment outside agriculture for the rural working class; the limited supply of council housing justified the continuation of the tied-accommodation system; and the low level of council rates was used as an argument against wage rises. As such, local government under the leadership of farmers, 'provided the essential infrastructure whereby the powerlessness of local agricultural workers has been maintained and a low-wage rural economy perpetuated' (Newby *et al.*, 1978, p. 274).

One thing that the agrarian elite could not control, however, was the attractiveness of the East Anglian countryside to ex-urban in-migrants. The trend of counterurbanization was just gathering speed as Newby *et al.* completed their study and they noted the arrival of middle class in-migrants pursuing the rural idyll to be 'the most recent threat to the political domination of landowners and farmers' (p. 248). Yet, forced to speculate about the likely effect of this challenge, Newby *et al.* argued that the in-migrants and the agrarian elite shared a common interest as property owners and that both had an interest in conserving the countryside. Thus, whilst the incomers had a more 'rigid' idea of conservation, they concluded that conflict was likely to be limited to localized disputes over 'uprooted hedgerows' and 'diverted footpaths' and that at the larger scale the in-migrants would be incorporated into the existing elite and therefore not destabilize the oligarchy of farmers and landowners. As the next chapter will discuss, this prediction proved to be mistaken.

Conclusion

The narrative presented in this chapter provides a clear demonstration that far from the British countryside being a stable, settled and largely 'apolitical' space until the disruption of recent restructuring, the local power structures of rural areas have been in a state of flux and dynamism throughout the twentieth century. At one level, changes to the rural power structure have been provoked by external events, albeit not necessarily immediately. The formalization of local government at the end of the 19th century and the First and Second World Wars all were destabilizing events for rural society that undermined elements of the existing political order and created opportunities for new leaders to emerge. The re-organization of local government in 1974 had a similarly impact, producing a potential for change in two ways. Firstly, the new councils created in the re-organization disregarded the existing spatial division of power. New county councils that amalgamated old county councils and county borough councils, and new district councils that merged together rural district councils, urban district councils and municipal boroughs, brought together on one authority the agrarian elites that had dominated rural councils and the business elites that had dominated urban councils. Although they shared a common conservative persuasion and hence assumed formal control of most of the new councils, they had to adapt to working together and farmers could no longer rely on monopolizing key positions (Woods, 1997). Secondly, local government in many rural areas became party-politicized overnight. In part anticipating the presence on the new councils of urban Labour councillors, Conservative Central Office issued an edict that members should contest the elections as party candidates, whereas many had previously sat as independents. The new partisanship, the disruption of change and the abolition of the aldermanic bench on the county council all encouraged many older councillors to retire and a new breed of party councillors who regarded local government as a step to a political career began to appear. Most significantly, perhaps, the explicit identification of the old rural elite with the Conservative party made them vulnerable to protest voting informed by national politics, as became clear in the 1980s and 1990s as the Alliance/Liberal Democrats made substantial gains on many rural councils.

At another level, changes within rural society have forced changes in local political leadership. The modernization of agriculture directly weakened the position of farmers as leaders of rural communities, whilst the closure of village services has reduced the number of visible rural 'political occupations'. The increased social and physical mobility of rural residents has undermined the discourse of the organic, agricultural community and the power of paternalism. Perhaps most crucially, the large-scale influx of in-migrants into rural areas in the last quarter of the twentieth century has introduced a numerous new social group into the rural population, many of whom have been prepared to challenge the authority of the traditional elite.

The balance of local power in rural Britain therefore followed a trajectory through the twentieth century from domination by a single, clearly-defined and exclusive elite of landowners at the start of the century; to a broader alliance in the mid part of the century in which power was shared between an agrarian elite that was pre-eminent in rural communities, a business elite that monopolized the government of small towns, and the remaining landed gentry who continued to exert influence at a county level; to a far more complex and fractious situation at the end of the century. Yet, as the next chapter describes, the local power structure of most rural areas at the start of the twenty-first century is still essentially elitist and still founded on the three dimensions of resource power, discursive power and associational power.

Chapter 3

Contemporary Rural Elites

Introduction

As Newby *et al.* (1978) were completing their study of East Anglia in the mid 1970s they glimpsed the beginnings of a process that would transform the local politics of rural Britain. The depopulation of the countryside that had persisted since the mid nineteenth century was reversed as middle class households started to move out of towns and cities into rural areas. Between 1981 and 1991 the population of 'mainly rural' districts increased by 469,000 people, or 7.6 per cent; whilst in the twelve months before the 1981 census, over 100,000 people were recorded to have moved directly from urban to rural areas. Urban to rural migration is motivated by a wide range of different factors that will vary between individuals, but for many migrants the decision to move to the country was strongly influenced by the 'rural idyll'. Having invested both financially and emotionally in this idealized concept, in-migrants were prepared to mobilize politically to defend their dream, including mounting challenges to the existing rural leadership where this was deemed to be necessary.

The first migrants, however, tended to be drawn from the more affluent parts of the middle classes, and as discussed in the previous chapter, Newby *et al.* (1978) concluded that they posed little threat to the dominance of the agricultural and landowning elite:

> Whatever the subjective differences perceived by farmers and newcomers over such matters as the environmental consequences of farming practices these have had little effect upon their common pursuit of local political policies. In this sense the newcomers have not presented the threat to the local political system that the frequent skirmishes over hedgerows and footpaths might indicate. By and large the former oligarchy of farmers and landowners has experienced few difficulties in incorporating such newcomers into the prevailing political practice. (Newby *et al.*, 1978, p. 275).

Yet, a mere twelve years later, Cloke (1990) reported that the in-migrant middle classes were seen as being influential in rural politics, and that they were seen to be in competition with the agricultural elites. So what had happened? The difference between the two statements reveals both a change in the nature of counterurbanization during the 1980s, and a change in the way in which rural researchers approached the question of class. With respect to the former, as counterurbanization intensified in the 1980s, so it also became more inclusive, embracing members of the lower middle classes whose socio-economic background made them less obvious allies of the traditional rural elite. Moreover, as in-migration to rural areas began to be accommodated more in large, new, purpose-built housing estates than the through the purchase of individual existing rural properties, so the newcomers were often physically separated from the 'old' rural community, militating against integration. With respect to the latter factor, the 1980s saw the development of a critique of Newby *et al.*'s model of rural class relations. Instead of focusing on property ownership as the basis of class relations, as Newby *et al.* had done, critics argued that occupational and lifestyle characteristics were more important in defining class divisions and focused on the transformation of the spatial division of labour in their analysis of rural class recomposition. As such, Cloke (1990) criticized the equation of 'incomer' with 'middle class' and proposed that the middle class was in fact comprised of a number of different class fractions involved in intra-class conflicts that had 'become important both per se and because such conflicts can change the shape of classes by strengthening or undermining their organization' (p. 308).

Particularly significantly among the fractions of the middle class in rural areas is the so-called 'new middle class' or 'service class'. Defined as professional, administrative and managerial workers who are relatively highly placed in the capitalist social hierarchy, members of the service class are not business owners but rather provide specialist inputs into the servicing of capitalist accumulation, in both the private and public sectors. The service class hence includes occupations such as accountants, human resource managers and other middle managers, marketing personnel, teachers, engineers, local government officers, junior civil servants and junior National Health Service doctors. According to Urry (1995), the service class is characterized by 'rapid numerical growth, high levels of educational credentials, a considerable degree of autonomy and discretion at work, reasonably high incomes (but lower than its cultural capital), opportunities for promotion between enterprises and relative residential freedom' (p. 209). The last two characteristics in particular have enabled the service class to become a key group in urban to rural migration, associated in part with a growth of service class jobs in urban fringe and small town locations. Furthermore, Thrift (1989) argued that service class culture values the traditional ideals of the 'rural idyll' and that therefore service class lifestyles seek to consume the rural idyll through tourism and day-trips, the purchase of particular fashions and houseware, and membership of organizations such as the National Trust and the Royal Society for the Protection of Birds (RSPB). Full pursuit of this lifestyle required certain types of housing located within reach of particular 'theatres of consumption', typically small towns with historic centres.

The service class has therefore both established a significant presence in rural areas and has become increasingly influential in rural politics. Cloke (1990) suggested that the service class had become influential in the Conservative and Liberal Democrat parties (and, it might now be argued, in New Labour), that service class cultural values (such as the preservation of traditional landscapes and of historic buildings) had become dominant in policy discourses, and, notably, that the service class had increasing representation in rural local government.

The politicization of the service class was assisted by the political resources that they possess and by their ability to network. Many service class occupations require skills that are also valuable in politics. Thus, members of the service class tend to be articulate, knowledgeable and well educated with good communication and organizational skills. In-migrant service class members may have been involved in local politics in their previous, urban, homes, and therefore bring political experience with them. Moreover, the substantial numbers of in-migrants who have retired from service class jobs to the country have time to participate in local community active and politics; whilst service class workplaces, such as schools, hospitals and offices, provide good opportunities for networking.

The actual mobilization of the service class, however, has frequently been triggered by particular local disputes that have provoked them to act in order to defend their investment in the rural idyll. It is therefore not uncommon for such trigger points to be provided by conflicts over footpath access – compromising the enjoyment of the countryside – new housing development – disrupting a view or expanding a village beyond what is perceived as an appropriate size – or street-lighting. As Harper (1988) observed, 'individuals thus compete for positions on the Parish Council in order to defend their image of the settlement'. Such disputes have often brought service class in-migrants into direct conflict with farmers and other members of the traditional power bloc. As new residents have felt that their interests and communities are not represented by the existing local leadership, and as they have found their ability to join existing elite networks limited by a lack of the right 'cultural capital', so they have become involved in local politics themselves, as two accounts from Somerset demonstrate:

> We had moved into a very large new housing estate area and the parish council really existed in the old country village of the locality. And all the councillors lived in the old village area and Galmington at that stage had not become at all political or motivated in any way, people were still moving in and they hadn't got themselves interested in local affairs. And what happened was that Galmington was putting forward some money to Trull Parish Council and getting none of the benefits, it was all being spent within the old village part ... Galmington had to get a number of parish councillors on to that parish council in order to get proper representation. (Interview – I413 District Councillor).

> There'd been planning permission difficulties when we wanted a new church and we were aware that things were being said at parish council meetings about the church which were perhaps unhelpful, and felt it would be a good idea to have someone on the parish council so that we couldn't be talked about behind our backs really. (Interview – I15 County Councillor).

In-migrants who have become politically active in such circumstances have not necessarily welcomed by their fellow councillors. They have often found themselves marginalized by the exclusivity of the existing elite and by the reproduction of discourses of localism and conservatism. Yet, at the same time, the professional skills of service class councillors have enabled them to work effectively on councils and committees such that their influence rapidly increased:

> When I first joined the village hall committee in '77, it was perfectly appalling, it really was, because it was run by this district councillor, who's now dead, who virtually built it, and the secretary also had virtually built it, and they were in cahoots those two. I mean, they had a committee but they never did anything, well, they never did anything the committee said. That's not strictly true, but they just carried on their own merry way, they couldn't care less what anybody else said. And as I say, if you were an incomer, my God, you were sort of less than nothing, basically, you didn't know what the hell you were talking about. (Interview – I120 Former District Councillor).

> The values which the parish council seem to have are more towards trying to save money than investing in the future. Because they have that policy of saving money and keeping the costs of user facilities like the village hall to a minimum, over the last twenty years they've not kept up the maintenance ... And all this is due to neglect by the existing establishment in the village. Now of course when people like me come in, I became a parish councillor first of all, I then became a county councillor; obviously you're a fast learner ... it's not long before you know they can no longer bluff you with what went on twenty years ago. (Interview – I6 County Councillor).

There is hence significant evidence that the service class have become an important group in the local politics of rural Britain, and Cloke and Goodwin (1992) went as far as to suggest that the influence of service class fractions 'is now such that they can be regarded as emergent historic blocs in some rural areas' (p. 328), able to pursue their own sectional interests in the regulation of the rural society and economy. Murdoch and Marsden (1994), on the other hand, dispute the existence of a service class hegemony in Buckinghamshire (an area that might conceivably be covered by Cloke and Goodwin's assertion), but do acknowledge that there are 'middle class cultures' in the county that are becoming hegemonic. Certainly, it would be erroneous to claim that the service class has become *the* dominant force in rural politics, or indeed, that the service class is able to organize itself in as coherent and discrete a way as the traditional rural elites. Neither is it the case that the established elites have disappeared. Farmers are still over-represented on many rural councils, and especially on parish councils. The business community still has a strong presence on town councils and in the realms of non-elected local government. Even the aristocracy and landed gentry are still active, albeit largely in positions such as Lord Lieutenant or High Sheriff or as magistrates, or through the quiet, informal, exercise of influence. Paxman (1991) observed that when the Duke of Richmond called a meeting of local business leaders to discuss the West Sussex Structure Plan, they all turned up, and quotes the Duke's remark that, 'I've proved there is still a role for people like me who live in houses like this ... if they choose to exercise it' (Paxman, 1991, p. 45).

The remainder of this chapter explores this contemporary rural power structure through the case study of western Somerset in the mid 1990s. It describes the elite networks that were active in local politics, and examines the ways in which they maintained their status and exercised influence.

Contemporary Local Politics in Somerset

The county of Somerset, in south-west England, has had a fairly typical experience of rural restructuring in the late twentieth century. Agriculture has declined in its economic importance, although it remains significant for many smaller rural communities. The service sector, manufacturing and tourism – particularly along the coast and in west Somerset – are now the major employers. The county has also experienced significant population growth through in-migration, especially around the towns of Taunton and Yeovil and in areas accessible to the M5 motorway or A303 trunk road. It is, however, beyond the commuter belt of London and therefore has not been reconstituted as such an exclusively middle class space as has the south-east of England (although commuting to Bristol is significant in the north of the county). Much of the migration to Somerset has been associated with employment relocation, with lifestyle change or with retirement, and the service class are well represented among in-migrants, not least because of the growth of Taunton as a regional service centre.

The analysis in this section focuses in particular on the two districts of Taunton Deane and West Somerset that comprise the western third of the county. Taunton Deane, with a population of 93,958 in 1991, is dominated by the county town of Taunton and its suburban communities of Bishops Hull, Comeytrowe, Norton Fitzwarren, Staplegrove and Trull (combined population, 56,414), and bisected by the M5, access to which helped to attract in-migrants who added 9.5 per cent to the district's population during the 1980s. In-migration has supported the expansion not only of Taunton, but also the smaller town of Wellington (11,287 population) and the large villages of Bishops Lydeard, Creech St Michael, Milverton, North Curry and Wiveliscombe. In contrast, the fringes of the district touch the edges of the Quantock Hills, Brendon Hills, Blackdown Hills and Somerset Levels and are more sparsely populated.

West Somerset is larger than Taunton Deane in area, but has a much smaller population of just 34,356 in 1991. The largest town is the seaside resort of Minehead (11,883 population), with other centres of population including Watchet (3,430), Williton (2,461) and Dulverton (1,406). Over half of the area of West Somerset lies within Exmoor National Park and large parts of the district are sparsely population upland on Exmoor and the Quantock Hills. The single largest local employer is the Hinkley Point Nuclear Power Station in the far eastern corner of the district.

Historically, both Taunton Deane and West Somerset have been Conservative in their political allegiance. The two districts are split between the Taunton and Bridgwater parliamentary constituencies, both of which elected Conservatives at every election between 1950 and 1992. Both Taunton Deane Borough Council and Somerset County Council were Conservative controlled for a decade after local government re-organization in 1974, but the Liberal Democrats gained support from the mid 1980s onwards. The Liberal Democrats took control of Taunton Deane Borough Council in 1991 and held on to the council in 1995 (Table 3.1) (the council went to 'no overall control' in 1999). On the county council, the Liberal-SDP Alliance formed a minority administration between 1985 and 1989, when the Conservatives regained control. In 1993, the Liberal Democrats were elected to a majority on Somerset County Council, winning nine of the 17 wards in Taunton Deane and West Somerset. The advance of the Liberal Democrats was also marked by the capture of the Taunton constituency in the 1997 general election (regained by the Conservatives in 2001). Independents formed a comfortable majority on West Somerset District until 1995, when both Labour and the Conservatives gained seats (Table 3.2). Of the 12 Independents elected to West Somerset District Council in 1995, four were members of the Conservative party.

Table 3.1 Composition of Taunton Deane Borough Council, 1973-1995

Year	Cons	LD/Lib	Lab	Ind
1973	22	0	13	13
1976	26	0	11	11
1979	26	0	14	9
1983	32	1	10	6
1987	28	15	7	3
1991	13	29	7	4
1995	14	29	7	3

Table 3.2 Composition of West Somerset District Council, 1973-1995

Year	Cons	LD/Lib	Lab	Ind
1973	3	1	1	27
1976	5	1	0	26
1979	3	1	0	28
1983	4	1	0	27
1987	6	0	0	26
1991	5	1	3	23
1995	9	2	8	13

Political pluralism, however, did not mean that there was no longer a concentration of power in western Somerset. The members of the two district councils and the county council were mostly male and disproportionately drawn from middle class backgrounds. Moreover, there was a significant over-representation of certain occupational groups, including farmers, employers and managers, and 'education professionals' – teachers, college lecturers and 'education consultants' – broadly reflecting the trend discussed in the previous chapter (Table 3.3).

Table 3.3 Representation of key groups on local councils, 1996

	Farmers	Employers & Managers	Education Professionals
Somerset CC	8.9%	10.7%	15.8%
Somerset workforce	1.1%	11.3%	3.7%
Taunton Deane BC	3.8%	13.3%	11.3%
Taunton D workforce	1.7%	11.1%	3.7%
West Somerset DC	11.1%	22.0%	12.8%
W Somerset workforce	3.7%	10.0%	3.4%

Sources: Survey of councillors, council information, census data.

The over-representation of these groups is in part due to their resource power. Wealth was identified as important by some interviewees who attributed power in the locality to businesses, particularly builders and developers. However such suggestions of covert business power alluded more to a perceived indirect influence on local governance than to formal participation in elected local government. On a broader scale, personal, cultural and social capital were of more significance than material capital in positioning individuals in the local power structure. Time availability, communication skills, confidence, motivation and knowledge were all proposed as factors that either permitted participation or formed barriers to participation:

> There's a lot of councillors now that are excluded from being in the swim because they can't give up the time, you know, they go to meetings, they haven't read the reports, they haven't read the papers, so they don't know very much, so there are a few now who are being covertly excluded because they haven't ... you know, they've got full time jobs, maybe a family, and from the point of view of getting people to be candidates, this is an unfortunate thing that has developed. (Interview -- 127 Taunton Deane Borough Councillor).

> There was a Tory councillor who joined the council at the same time as me in '87, her husband worked on shift and her children were, oh, seven or eight or something like that, and after a short period, after a short time, she said 'I'm running out of babysitting points', and after four years she stood down. It didn't, it didn't fit together. (Interview – 129 Taunton Deane Borough Councillor)

Clearly, however, such personal attributes are not enough on their own to bring influence. There are individuals who have these personal resources but have been thwarted in their attempts to be elected or appointed to positions of local leadership because they do not have the appropriate networks of contacts. This is only in part about 'political occupations' (see chapter two), although on West Somerset District Council, which was still largely non-partisan in the mid 1990s, councillors included a village garage owner, a village shopkeeper, a retired postmaster, a retired community worker, a retired butcher, a baker, two nurses and three teachers. More significant, however, were the contacts that made an individual part of an elite network and therefore permitted access to the flows of recruitment, patronage and influence through such networks.

Elite Networks in Somerset

Elite networks are clusters of individuals who hold positions of power and influence and are connected by social, personal or family ties. The term is used deliberately as a critique of conventional elite theory and emphasizes the notion that elites are neither discrete nor homogeneous, but rather are composed of a network of social and professional relations, both between members of the elite, and extending outside to individuals in the non-elite (see also Woods, 1998b). Elite networks are by nature fluid and dynamic with no fixed boundaries. As such, identifying elite networks is a necessarily imprecise process. For the empirical case study that is discussed here, the elite networks were identified by collating and mapping information about ties between individuals who held office in local governance or who were politically active in the locality, drawing on a questionnaire survey, local newspaper reports and library and archival sources. In particular, emphasis was placed on common membership of clubs and societies, committees and churches; shared workplaces; residential proximity and family ties. Common membership of political parties was not included. This method is inevitably fraught with difficulties, says nothing about the frequency or quality of interactions, and produces an incomplete result. A note of caution must therefore be attached to the description of networks in this section, but they nonetheless present a workable approximation of the social topography of the local power structure in the area.

What this approach reveals is that in western Somerset in the mid 1990s, a considerable number of local political activists were linked through social and professional relations into an elite network, which can be further divided into three 'cores', each with a distinctive social and political character. The first core was in effect the continuation of the 'traditional rural establishment' – middle class, politically Conservative with a strong agricultural bias. Amongst the individuals included in this network were the Lord Lieutenant, the chair of the county council, and a number of present and former councillors and holders of non-elected governance positions, who were linked together through family and professional contacts, and through their involvement in organizations such as the Royal Bath and West Agricultural Society, the Red Cross (Figure 3.1), stag- and fox-hunts, and the National Farmers Union (Figure 3.2).

The second core consisted largely of local businesspeople, most of whom were Conservative in their politics. They were closely linked through organizations such as the Chambers of Commerce, Rotary Clubs and Masonic lodges in Taunton, Minehead and Wellington, the Taunton Court Leet, Taunton Soroptomists and the

Contemporary Rural Elites

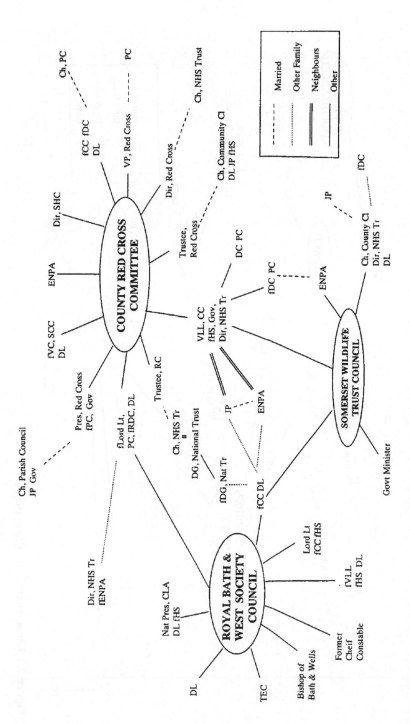

Figure 3.1 The traditional county establishment network

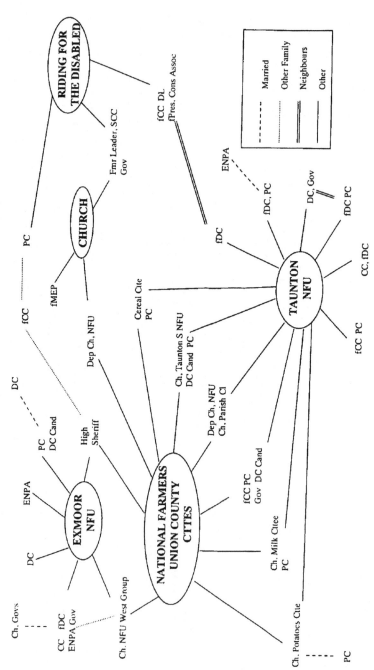

Figure 3.2 The agricultural elite network

Minehead Gala Committee, and included a number of senior figures in non-elected local governance, such as the chairs of the Somerset Health Commission, the Taunton and Somerset NHS Trust and the Training and Enterprise Council, as well as the district coroner and the Conservative leader on Taunton Deane Borough Council.

Finally, a third, less distinct core, was based on connections through social and voluntary organizations in Taunton and service sector work places. This 'emerging liberal elite' is also less socially or politically homogeneous, but includes a significant number of 'service class' in-migrants, who tend to be Liberal Democrat in their political affiliations, and include the Leader of Taunton Deane Borough Council, the chair of the Taunton Magistrates Bench and a number of district and county councillors (Figure 3.3). They are, therefore, the group that is seen by some as replacing the previous hegemony of the business elite in Taunton:

> There are a lot of county council employees who are members of the borough council ... and then I think the utilities, electricity, gas, BT and whatever, are another provider of candidates and councillors. So, the utilities and the ... county council in the main, have significantly replaced and probably exceeded the previous influence of the business community. (Interview – 150 Taunton Deane Borough Councillor).

Simply identifying the existence of such networks, however, does not reveal anything about their consequences for the practice of local politics and governance. A group of councillors may all be members of an organization, attend meetings, and yet this may have no effect on their work as councillors. Hence we also need to look at the ways in which elite networks are used for recruitment, patronage and the exercise of influence.

Recruitment

The use of social contacts is highly important in the initial stages of recruiting people into political activity. Social networks provide an important means of recruiting members for all political parties, but there is also a second stage of recruitment when people who have shown a political interest are then recruited into political activity – a process that may be the beginning of recruitment into the local elite. This is perhaps most commonly performed in terms of individuals being approached and asked to consider standing as a candidate for election to one of the councils. Except in a few favourable localities, all the parties have experienced a dearth of willing potential candidates putting themselves forward for election, and the fielding of a respectable slate of candidates is more often the product of active recruitment. As such many councillors report that they first stood for election as a result of being asked to do so by a party or someone they knew:

> For twelve years I wasn't involved in politics at all, just single issue items, and really it was through social contacts that I was approached and asked to stand in eighty-seven. (Interview – 142 Taunton Deane Borough Councillor).

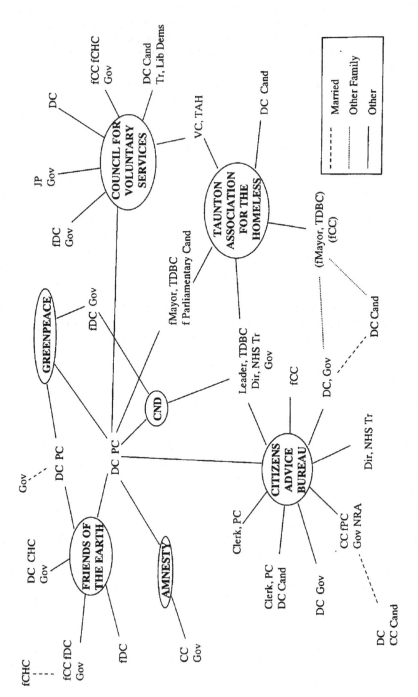

Figure 3.3 The emerging 'Liberal' elite network

> One of our local, erm, Tory members asked me if I would like to stand, and she gave me a booklet, at the time I was just flying out to Grenada ... so I took the booklet with me, read it, and thought 'Yes, I would like to do it'. Not, by any means, from a political angle at all ... that ... you know... I'm really not into politics whatever. I suppose I've always been a bit blue, but I'm not, you know, I don't consider the politics side of it ... I was really more interested in what was going on around in our villages and towns. (Interview – 1263 West Somerset District Councillor).

> I was then thought to be a suitable person to represent this area on the county council, *they* not being very happy with the person who was at that time doing it. In fact I would not have liked to have opposed this chap ... but extraordinarily the old chap died. And that was that, you know. So they were then looking obviously for a candidate and it was put to me that I should stand. (Interview – 1190 Former County Councillor).

The networks which existed between political actors and the way in which they were used in recruiting candidates were also evident on the nomination papers of candidates. In the May 1995 elections to Taunton Deane Borough Council, 41 candidates had their nomination papers signed by other candidates. For party political candidates, the list of nominations may simply reflect the attendance at the relevant branch meeting, and imply nothing except that there is only a small core of activists for the party in that ward. However, in non-partisan elections, the collection of nominations has to depend far more heavily on social networks. This becomes significant in parish council elections, which tend not to be contested by political parties, and where again there can be a severe shortage of self-proposing candidates. In order to ensure that candidates are found, individuals may take it upon themselves to recruit from among their own social network, or alternatively, a group of people who know each other as neighbours or friends may reach a collective decision to put themselves forward, as happened in one village in Taunton Deane:

> A group of younger people said, 'We'll give the village an opportunity to elect somebody younger'. Not that there was any great dissatisfaction with the parish council, it was just that they thought ... I believe I'm right in saying ... and not as an organized group, certainly much less any party political intention behind it, they, you know, just put their names forward, no doubt having talked about it to each other. (Interview – 121 Taunton Deane Borough Councillor).

A more subtle way in which social networks have had the effect of recruiting political actors is by what might be termed, 'activating by example'. In other words, there is no deliberate attempt to recruit an individual through a social network, but knowing someone active in local politics through a social network encourages an individual to become involved themselves. Again, this kind of experience features in the life-histories told by some councillors:

> At parish council meetings was the district councillor who was exceedingly good, very well motivated, he cared about the people and about local services, and the county councillor used to come along to the meetings as well, and I was impressed

by this district councillor and county councillor, by the way they really cared, fought hard for the things that people wanted, and were prepared, if necessary, to get their hands dirty to get things done. And that impressed me and I thought, 'I could do that, that's the sort of councillor I could be.' (Interview – I15 County Councillor).

In my very early teens I was friends with the now mayor, her son, they were active Labour people ... so from my earliest days I was aware of party politics. I went and stuck up skittles for the Labour Party. I never joined them, vaguely worked for them ... It evolved through that in my latter teens into supporting the Liberals. (Interview – I255 Taunton Deane Borough Councillor).

Social networks are thus seen as contributing to raising people's political awareness and later providing opportunities by which individuals might be recruited into political activity.

Patronage

Once recruited into political activity, the importance of an individual's social network does not diminish. Rather the new contacts developed through political activity incorporate an individual into new social networks and begin to open up new opportunities as to how their political career might develop. I am referring here to *patronage*, which I distinguish from *recruitment* as being concerned with appointing people already active in politics as opposed to recruiting people into politics. The areas of local governance most commonly linked with patronage are the various non-elected boards such as health authorities and NHS Trusts, institutions that are sometimes defended as widening the range of people involved in local governance:

I do believe that, out there, there are hundreds of people with considerable talents, limited time, who would genuinely like to make a contribution to making the wheels turn, but who don't fancy the idea of every four or five years going around and banging on doors saying 'will you vote for me'. There's nothing dishonourable about that, and I think that quangos gave us the opportunity to tap that considerable reservoir of talent and goodwill. (Interview – I156 Former Health Authority Chair).

The problem that arises with this approach, however, is that the people or person responsible for the appointments has only limited knowledge of what people there are 'out there', willing or wanting to 'make a contribution'. Members of their own social and professional networks are the most obvious candidates, and attempts to widen the field frequently mean little more than tapping into the social networks of associates:

You touch on a problem when you're appointing to these bodies, because for everybody you know for good or ill, there are a hundred and one very worthy people you haven't got to hear about. And you have to canvass around a bit, and ask people whose judgement you respect 'do you know anybody who might be worth looking at?' and you gradually build up. (Interview – I156 Former Health Authority Chair).

This radial process means that people who are active in local politics, people who are known to those with the power to appoint, or those consulted by the appointing authority, are more likely to be appointed than someone with equivalent skills but outside those networks. The experience of two members of appointed bodies in Somerset followed this pattern:

> It was really through my links with Angela Sneddon, who was then the leader of the Labour group on the county council, who was somebody I met at various meetings ... As leader of the Labour group on the county council when there was a vacancy for a nomination on behalf of the county council to the community health council she was asked if she had a name to put forward ... And she asked me if I was interested in it and she knew my background. (Interview – 1410 Community Health Council Member).

> I find that with my local contacts when I was chucked off the council it meant that I still had contact to a certain extent, I still knew what was going on. And, from the point of view of that, I'm now chair of a district and county arts committee. (Interview – 1120 Former Taunton Deane Borough Councillor).

This kind of patronage leads to the concentration of political and governmental roles with a fairly small group of people. Over a third of Taunton Deane Councillors in May 1996 also held some other local governance office – as a school governor, magistrate or member of an appointed body. A third were also members of other councils, whilst nearly half held offices in non-political organizations. At the same time, there was also a small group of councillors who did not have any outside activities. One interviewee suggested that there are two types of councillor, those who 'become a councillor and don't do anything else', and those who are councillors 'with a lot of strings'; but, this can be expanded to produce a rough typology of five types of local political activists:

1. *County leaders*, who are active in a number of (often county-wide) organizations at a senior level and who also tend to hold senior positions in local councils or non-elected local government.

2. *Community activists*, who are active in a number of organizations, but not necessarily at a senior level.

3. *Professional politicians*, who are very active in political parties and/or local government, but who do not have many interests outside this narrow field. Some have taken early retirement to concentrate on political activity.

4. *Specialists*, who are very active with another narrow field, such as hunting, the environment or the arts.

5. *Representatives*, who may serve as councillors, but not in a senior position, and who do not have many outside activities.

Each of these types of activists can be found in the elite networks in Somerset. The 'county leaders' and 'community activists' tend to be the key figures in the networks, with their numerous contacts bridging various fields of interest. The participation of councillors and other politicians in non-political organizations, such as hunts or a hospital league of friends is important in bringing 'specialists' into a network, whilst councillors with no outside interests are least likely to be integrated into a network, a position which might be self-reproducing as they are then less likely to be the recipients of patronage. The 'county leaders' tended to be from landowning or senior business backgrounds, and were mostly to be found in the 'traditional establishment'; the 'professional politicians' in contrast, tended to be from the professional middle classes, and were more prominent in the 'emerging liberal elite' – indeed their narrow but intensive political activity may be viewed as a strategy for matching the benefits which 'county leaders' accrue from flexible time, wealth and patronage. Perhaps most significantly, the elite networks provided channels of communication and interaction between the different types of political actors – which leads to the third way in which networks are used, for the exercise of influence.

Influence

The exercise of influence through elite networks is more ambiguous than the use of networks for recruitment or patronage, not because the consequences are less visible, but because there is a multitude of ways in which it is done. At one extreme there are fairly direct, overt, ways of trying to influence decision-makers in private meetings – the old 'smoke filled rooms' metaphor. However, perhaps more important are more subtle forms of influence, whereby knowing someone through social contacts means that a decision-maker is more informed about a particular position or argument than another, or is more inclined to be sympathetic to a particular case. This kind of influence was seen as working between councillors and council officers, and between junior or opposition councillors and council leaders:

> On your day to day contact with the officers, I think if you get to know them socially then you're far more likely to get them on your side and if they don't want to do something and you do, they're in a position that it could get slowed down, that they do it at the pace they want to do it, but if you've got a friendly relationship, they're far more likely to do it at the speed that you want to do it. (Interview – I81 Taunton Deane Borough Councillor).

> Well, Jeff Horsley, he's the leader, he and I have been involved in local politics for a long time, CND as well, many, many years ... yeah, we have a good measure of things that I see eye to eye with them, we work together. (Interview – I41 Independent Councillor).

Thus the stories told were of networks being used to gain support for an idea in order to accumulate more power or influence than an individual's actual office allows. This is an interesting observation when compared to commonly held perceptions of power and influence, challenging even the power that is popularly attributed to senior figures such as council leaders:

> I wouldn't over-estimate my role. I'm only one out of fifty three votes, when it comes to actually, say the budget, I have to carry, I have to enjoy the full support of my colleagues within the group for me to exercise that influence. (Interview – I400 Council Leader).

What is being described here is the mobilization of associational power. To exercise influence in the council chamber a councillor must secure the support of a majority of colleagues – a process that involves both formal party political structures and social networks. In this context the comment of one town councillor that 'in Minehead we work as a team', turns from being at first glance an attempt to create an impression of consensus, to being a statement of pragmatic necessity. Moreover, the enrolment of actors involves valuing their potential influence within particular contexts. Each individual's own potential influence is essentially restricted by their office, their interests and their place in the elite network:

> I think 'power' and 'influence' flow within circles or areas of interest. John Meikle and Jefferson Horsley have influence in the area of Taunton Deane Borough Council. Sir Walter Luttrell and Lady Gass both have influence on 'countryside' matters in the Quantocks/Exmoor. I don't know who has the greatest influence in farming or business in the TD & WS areas. Many people in the arts share influence, no dominating name comes to mind. The Bishop of Taunton in Church affairs. Chief Exec., Somerset CC, Brian Tanner, has quite wide and general influence. (Questionnaire – Q9 County Councillor).

> There isn't anybody with power as such, because, don't forget, none of us have got power outside the council chamber, you know, as chairman of the town council, once I leave the town hall I've got no power whatsoever. (Interview – I78 West Somerset District Councillor).

The relevance of elite networks in this process is that social contacts are mobilized by individuals to gain support from actors active in other fields than their own with whom they have no formal official relations, but whose influence may be important in furthering a particular cause.

The Spatiality of Elite Networks

The exclusivity of elite networks is produced by their function in policing access to the 'backstages' of local politics, where the processes of recruitment, patronage and influence are performed. To become part of the network, or to improve one's position in the network, an individual has to belong to the right organizations, work in the right places, go to the right social events – or find other ways of forming and enacting social relations with the right people. As elite networks are therefore constructed through interactions that take place in particular spaces, the organization and operation of an elite network has a distinctive spatiality. In Somerset in the 1990s, the 'traditional rural elite' could be described as the 'county establishment' not just because of its participation in organizations that constructed themselves as 'county-wide', but because its members were drawn from across the county. The 'business elite' in contrast was more tightly concentrated on the three towns of Taunton, Minehead and Wellington, which were the commercial centres of the locality. This spatial compaction helped them historically to dominate local politics in those towns, but equally contributed to the loss of their direct influence in local government following the abolition of the compact borough and urban district councils in 1974. Thirdly, the 'emerging liberal elite' was rooted in Taunton and neighbouring villages that had experienced considerable recent growth, as befitted its incomer-middle class profile.

A further aspect of the elite networks' spatial identities are the literal 'meeting places' in which the networks are materialized. These vary in their spatial fixidity and their public accessibility. At one extreme many spaces of interaction such as society meetings, sports events or hunt meets are purely ephemeral, bringing together people in one place for only a short period of time. Often such meetings might occur in different locations each time, although others become identified with particular buildings that exist as permanent symbols of that organization's status in local society – for example Masonic Halls and churches. Indeed, it is interesting to note that whilst only 14 per cent of the population of Somerset claimed to attend church regularly (Brierley and Hiscock, 1993), at least 40 per cent of both Taunton Deane and West Somerset councillors in 1996 were regular church-goers, and a number of churches form significant 'elite spaces'. Thus, the congregation of St Mary's, Bishops Lydeard, included four district councillors, two magistrates and a clutch of parish councillors, and the congregation of St George's, Wilton, included one serving and three former district councillors – services providing an opportunity for interaction outside the formal spaces of local government.

Similarly, there are a number of workplaces with the potential for almost daily interaction between the political actors who are employed there. For example, the 1995 council elections were contested by all three of the proprietors of one Taunton residential home, with two being successful in getting elected to the district council. The partners of one Taunton accountancy firm, meanwhile, have included members of both the health authority and family health services authority, and a magistrate. In the public service sector, two district councillors were on the staff of St Augustine's School in Taunton, whilst the staff of Taunton School had over the decade from the mid 1980s to mid 1990s included three district councillors and two magistrates, and a further three councillors have been employed on the senior staff at Musgrove Park Hospital.

The distinction between churches and workplaces as elite spaces is in public accessibility. Churches and most societies are open to anyone to join and thus potentially offer a way into the elite network. Workplaces, however, are essentially restricted, as indeed are a number of societies that construct themselves as being exclusive:

> I mean there are certain societies where you attended when you were Mayor, women's societies, where they were quite happy to invite you as Mayor to give speeches and so on ... they would never invite you to be one of their members because you're a politician and you're a Liberal Democrat politician. (Interview – I6 Taunton Deane Borough Councillor).

Even more restricted is the kind of social interaction which occurs between elite members in entirely private spaces, and which may be constructed to be exclusive even to their political colleagues:

> There's a couple of my Tory friends, no Tory colleagues, drop friends put colleagues, who tend to pop round for G and Ts and the rest of it, but I'm not on that circuit because of my lifestyle. (Interview – I78 West Somerset District Councillor).

Such informal gatherings can be significant both in enabling small groups to discuss matters in private and as a space of discussion between politicians of opposing parties and between councillors and officers:

> Quite often after committee meetings or a full council meeting, a group of us will go to the local pub around the corner from the council chambers and have a drink, and you'll find that there's officers, Conservatives, Liberal Democrats, Labour, all socializing together, you know, when we're out of the chamber. (Interview – I81 Taunton Deane Borough Councillor).

Moreover, the spaces in which such interactions take place need not necessarily be located in Taunton Deane or West Somerset. As figure 3.4 shows, friendships formed as fellow students at public schools and university are still significant for the 'traditional rural elite' in particular, whilst a number of social relations in this network are mediated through London gentlemen's clubs such as Brook's and Cavalry and Guards, which like the Royal Bath and West Agricultural Society, provide an intersection between the traditional Somerset rural elite and national and regional elites.

The geography of elites that emerges from these stories is one which is at variance with the geography of local government as it is presented to the public. In the 1990s, both Somerset County Council and Taunton Deane Borough Council introduced initiatives to increase public participation in council proceedings, including public question times, the verbal presentation of petitions and advertising meetings. The implied message was that the council chamber is the locale of local power and allowing the public to participate was to give them a share of influence. However, if councillors themselves indicate that there are other, more restricted, spaces in which matters of local politics and government are discussed and even decided, then the public's access is only partial and 'elite spaces' remain significant in the political process. This can be explored further by following the operation of the elite networks in practice.

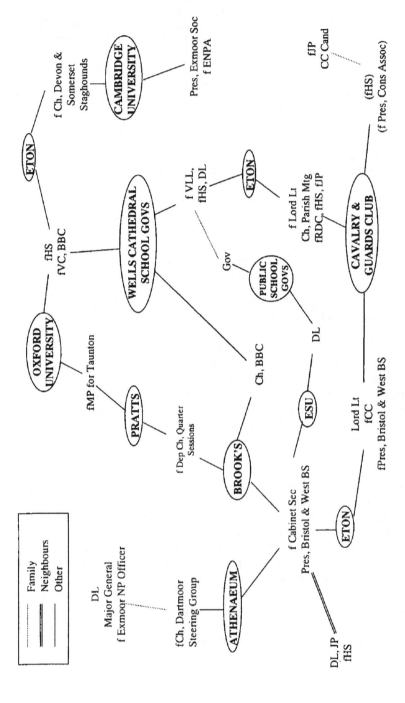

Figure 3.4 Networks through public schools and London clubs

Elite Networks in Practice

The Lieutenancy Dining Club

In the first week of October 1995, a small, 13-line item appeared in the columns of the *Somerset County Gazette*. Headed 'Dinner with the Lord Lieutenant', it read as follows:

> The Lord Lieutenant of Somerset, Colonel Sir John Wills Bt., the Deputy Lieutenants and honorary members of the Lieutenancy Dining Club dined on Thursday, October 3, at Hatch Court, Hatch Beauchamp. Sir John Wills presided, and General Sir Michael Rose and Brigadier Christopher Wolverson were the guests. (*Somerset County Gazette*, 6 June 1995).

This seemingly incongruous report was in fact a rare public glimpse of one of the most exclusive elite spaces in the county. As well as the Lord Lieutenant, Sir John Wills – a landowner, former county councillor and leading regional businessperson – the assembled Deputy Lieutenants included a former chair and a former vice-chair of the county council, the national president of the Country Landowners Association (later to become the Chairman of the Countryside Agency), and two board members of the Avalon NHS Trust – John White, the lord of the manor of Taunton, and Lady Gass, a county councillor and the only female Deputy Lieutenant. Although the identities of the 'honorary members' are not reported, they could be expected to have included the former Lord Lieutenant, Sir Walter Luttrell – a major landowner in western Somerset – and the High Sheriff, Roy Hewett (Figure 3.5). The venue too was significant, the imposing Bath-stone mansion had been home to two MPs for Taunton and had more recently hosted visits by senior Conservative politicians.

Collectively the members of the Lieutenancy Dining Club are the holders of some of the oldest local governance positions in the county. Lords Lieutenant were first appointed in 1457, charged with raising county militia; by the sixteenth century they had become the monarch's representative in the county and the main channel of communication between the government and the country; and by the eighteenth century they were responsible for the maintenance of order in the county, not just as commanders of the militia, but also as chief magistrate (justices of the peace themselves being of even greater antiquity) (Packett, 1975). Deputy Lieutenants emerged on an *ad hoc* basis in the sixteenth century to assist the Lord Lieutenant in raising the militia; whilst the origins of position of the High Sheriff appear to date from the Saxon era. Until recent centuries the sheriff was the major figure in local government, being responsible for collecting taxes, appointing members of parliament and administering the law (Astor, 1961).

With the introduction of institutional local government powers were removed from the Lieutenancy and Shrievalty, and they are most visible today in their ceremonial roles. The Lord Lieutenant is responsible for co-ordinating royal visits to the county, presenting medals and Queen's Awards, and presiding at certain civic ceremonies. The High Sheriff attends to visiting judiciary and presents

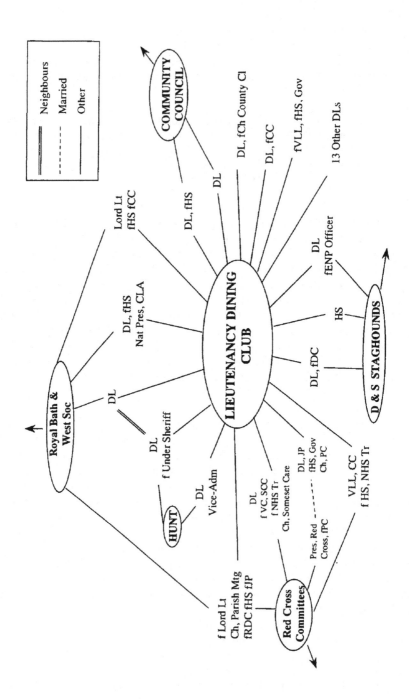

Figure 3.5 The Lieutenancy Dining Club network

gallantry awards (Astor, 1961; Packett, 1975). This visible ceremonial role has created a popular perception of the Lieutenancy and Shrievalty as anarchic, powerless, institutions. However, such a perception underestimates the practical role that they still play in local governance. The Lord Lieutenant (as Custos Rotulorum), is still chief magistrate, and as such is an *ex officio* member of the Magistrates' Courts Committee and chair of the County Advisory Committee on Justices of the Peace, which appoints magistrates (a proposal was briefly floated in 1995 that they should also oversee the appointment of independent members to police authorities). The High Sheriff is the official returning officer for the county; although this role is usually delegated to the Acting Returning Officer (a senior local government officer), the High Sheriff personally performed the role of Returning Officer for the Somerset and North Devon constituency in the 1994 European Elections. The High Sheriff is also technically responsible for summoning jurors and executing High Court writs, though these duties are delegated to the Under Sheriffs who are practising solicitors.

In recent years, High Sheriffs in Somerset have also taken a pro-active role in crime prevention in the county. The High Sheriff for 1994-5 organized a conference of the police, churches, youth organizations, social services, magistrates and probation officers to discuss strategies for crime prevention. Her predecessor was involved in an initiative aimed at reducing youth crime, whilst the High Sheriff for 1996 pledged himself to promoting a scheme working with schools to combat drug-related crime.

However, perhaps more important still is the informal influence attributed to the Lord Lieutenant, and especially the capacity for patronage within the Lieutenancy Dining Club:

> A Lord Lieutenant for a county has very little power, but he has considerable influence because he knows so many people and a lot of people are quite likely to have been to his soirees or whatever, and they don't want to offend him and from time to time they might even try to help him ... and so it's sort of very intangible. (Interview – I190 Deputy Lieutenant).

> I occasionally get asked will I do something, will I chair a forum on this that or the other, you see, and when I look into it I find, quite often, for example, somebody will have said 'well we have no idea who to invite to do this', and so they go and have a word with the Lord Lieutenant of the county, and he says, 'where will it function', you know, and then says 'well why don't you try so and so, he might be prepared to do it.' I mean, he's not bestowing patronage, he's dishing out dirty jobs for a start, but I mean, he would be a facilitator to put people in touch with other people. I mean, I happen to be a Deputy Lieutenant and I know the Lord Lieutenant has shunted one or two of these things in my direction. (Interview – I412 Deputy Lieutenant).

The influence that the Lord Lieutenant has is therefore perceived to stem wholly from his position in the elite network. As this influence rests on the Lord Lieutenant *knowing* people to nominate, the names he is likely to suggest are likely to be drawn from his social and professional network, with Deputy Lieutenants and former High Sheriffs prominent possibilities. From this perspective, the processes of appointing Lords Lieutenant, Deputy Lieutenants and High Sheriffs can be seen not as items of historical curiosity, but as events which may have a subsequent effect on the local governance of an area – and here a gap arises between the legal theory and practice. This is implied in a description of the appointment of the High Sheriff:

> It's open to anyone who owns property, although there is a tortuous appointment procedure ... At the end of each year the retiring High Sheriff draws up a list of suitable candidates. You can't apply, obviously. (Interview – I8 Former High Sheriff).

Thus, whilst technically all property owners in a county, with a few specific exceptions, are eligible to be nominated, the method of appointment – whereby the serving High Sheriff submits three names to the Queens Bench of the High Court of 'persons fit to serve as High Sheriff' (Astor, 1961, p. 15), one of whom will be appointed for the following year, and the other two in sequence – tends to ensure that High Sheriffs continue to drawn from within the elite network of which the retiring High Sheriff is a part (Figure 3.6).

A similar discrepancy occurs with the appointment of Deputy Lieutenants. Legally, the requirements of a Deputy Lieutenant are that they live in, or within seven miles of, the county and that they have rendered 'worthy service' in the armed forces or 'other such service as, in the opinion of the Secretary of State [for Defence], makes him suitable for appointment as a Deputy Lieutenant.' (Packett, 1975, p. 59). Thus one Deputy Lieutenant describes the office,

> It's a little honour given at a local level to those who've been of some service to the community, and at a tolerably high level, you know, and also because of its military history it is frequently though not universally given to senior officers who may retire in the county, therefore you'll find there's a splendid smattering of enormously distinguished senior officers on the list, though of course they're getting fewer and fewer in between. I mean most of the ones we've got are very ancient now, so you find the later ones that come up there are more and more misters appearing, but nevertheless they are people who usually have made some contribution. (Interview – I190 Deputy Lieutenant).

Inspection of the list of Deputy Lieutenants, however, proves the qualification of serving the community to be highly subjective. As the appointment is entirely the gift of the Lord Lieutenant it is their subjectivity that is important and the appointments reflect the biases of their elite network. Most are landowners and their contributions to the community tend to have been made through the kind of 'establishment' organizations through which the traditional rural elite operates. Of the Deputy Lieutenants for Somerset in 1996, three had been Conservative county councillors, one a Labour councillor, and one, Ralph Clark, the Liberal Democrat chair of the county council but also a former director of C. J. Clark and Sons shoe manufacturers and a member of one of the county's leading mercantile families.

Contemporary Rural Elites 71

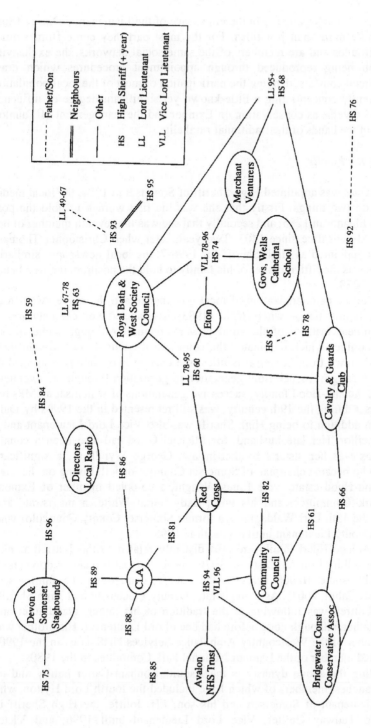

Figure 3.6 Connections between High Sheriffs

The select party gathered in the elite space of the Lieutenancy Dining Club, therefore, share more than just titles. For the most part they come from similar social backgrounds and are members of the same social networks, the exclusivity of their club being reproduced through appointment procedures which draw heavily on social contacts, denying the participation of most of the local population and consequently concentrating a little-known yet significant degree of influence on matters as diverse as conservation on Exmoor and the management of Taunton town centre in the hands of the traditional rural elite.

Keeping it in the Family

When Lady Gass was appointed High Sheriff of Somerset in 1994, the local media commented on two things. Firstly that she was the first woman to hold the post since Queen Eleanor in 1285, and secondly that she was the seventh member of her family to hold the office since 1403. Two years later when Christopher Thomas-Everard was appointed as High Sheriff for 1996-7, the local newspaper similarly noted that he was the fifth member of his family to hold the position, the first being appointed in 1258.

Despite the democratization of local governance and the move away from paternalism, family ties -- what Heald (1983) has labeled vertical networks -- remain prominent features of the rural power structure. Although some of the aristocratic families which dominated the county in the feudal era had sold their land and withdrawn from Somerset politics (see chapter two), others continued to be active in local governance from generation to generation through the twentieth century. The Acland-Hood family, successive generations of which sat as MPs for West Somerset during the 19th century, was still represented in the 1990s by Lady Gass, who in addition to being High Sheriff was also Vice Lord Lieutenant and a county councillor. Her late husband, Sir Michael Gass, had also been a county councillor, as was her sister's brother-in-law, George Wyndham, a significant landowner who became chairman of Somerset County Council. Living on the edge of the Acland-Hood estate were Ernest Wright, a co-opted member of Exmoor National Park Committee, and his wife, Lady Sarah White, a magistrate and daughter of the 13th Earl Waldegrave, a former Somerset County Councillor who served as a Deputy Lieutenant until his death in 1995.

The Acland-Hood family are also distantly related to the Luttrell family who, despite selling Dunster Castle in 1976, are the largest private landowners in Somerset with some 10,000 acres in the county, centred on the family seat Quantoxhead (Cahill, 2001). As a magistrate, Deputy Lieutenant and High Sheriff, Sir Walter Luttrell was following in the tradition of his father, grandfather and great-grandfather, although none before had been Lord Lieutenant. His brother was a board member of the Westcountry Ambulance Services NHS Trust in the 1990s and a co-opted member of the Exmoor National Park Committee in the 1980s.

Rivaling the above dynasty is the Asquith/Bonham-Carter family and its collateral branches, members of which have included the fourth Lord Hylton, who was Lord Lieutenant of Somerset, and his son, J.H. Joliffe, the High Sheriff in 1993; David Tudway Quilter, Vice Lord Lieutenant until 1996; and Victor

Bonham-Carter, a former member of the Exmoor National Park Committee and president of the Exmoor Society, among others. Additionally, family names such as Heathcote-Amory, Herbert, Lewin-Harris, Malet, Mitford-Slade, Trollope-Bellew and Vivian-Neal make multiple appearances in the records of twentieth century local government in western Somerset.

Political dynasties are not limited to the aristocracy. In the introduction to Griselda Fox's *Country Diary* (1978), her grandson describes the kitchen at Gerbeston Manor, including a table 'with one leg now supported on the minutes of the Rural District Council'. What he neglects to say is that not only was Griselda Fox chair of the rural district council, but her nephew was also a member, and that her husband was a county alderman and urban district councillor, her son a magistrate, and that the family had included three county councillors during the twentieth century – all largely based on their economic strength in Wellington where the family woollen mill was for decades the largest employer. Similarly, the Clark family, whose shoe manufacturing company is still one of the county's largest employers, has throughout the century provided numerous councillors and magistrates, including the chair of Somerset County Council in the mid 1990s; whilst four successive generations of one farming family have served as magistrates.

The explanation of such 'vertical networks' is not perhaps overt nepotism, but more a case of 'political families' playing a role in political societalization. In other words, growing up in a family that was active in local politics would have made an individual more politically aware and maybe inclined them more to get involved. As a relative of an established local political figure, they were also perhaps more likely to be approached by others and asked to stand for election, or be appointed to a particular position. This political societalization does not only apply to the gentry, as a lower middle class councillor demonstrates:

> My brother in law has been actively involved in local government and politics for about the last twenty years, and I started off by helping him with his election campaigns, firstly leaflet drops and then later on canvassing and I thought, when the Deane elections came up this particular year, you know, I know how a political campaign runs and you've got a bit about local politics, so I thought I'd stand and see what'd happen. I never expected to actually win the seat. (Interview – I81 Taunton Deane Borough Councillor).

Furthermore, nine married couples contested the 1995 district council elections in Taunton Deane and West Somerset, two of them where both partners were successful. Additionally, at least six councillors in the area in 1996 were married to school governors, and at least eight magistrates had partners who held office in local governance. This family involvement may again stem from living in a political environment, or may be in effect a mechanism for coping with the pressures of political activity – both partners being involved in the same activity might be a means of spending more time together, whilst having different interests is a way of gaining knowledge about other areas of local governance (see Barron *et al.*, 1991). The multiple participation of certain families in local governance need not have any paternalistic overtones attached to it, but the effect is nonetheless to concentrate local governance offices and potential influence in elite networks and to create further elite spaces (including households) to which the public is denied access, but where issues may be discussed, decisions taken and political knowledge accumulated.

Story 3: The Court Leet, Freemasonry and Taunton Business

On the first Friday in November each year, senior members of the business community in Taunton gather at the town's Municipal Buildings for the Law Day of the Court Leet. Watched by guests who include the Member of Parliament, the Mayor, the High Sheriff, the superintendent of police and the lord of the manor of Taunton, the thirty officers of the Court Leet are sworn into office for the following year. The ceremony is an ancient one. Dating from mediaeval times, the Court Leet was the original institution of local government in Taunton. As a Court of Record it had the same power in the Hundred of Taunton Deane as the High Sheriff had in the county; it was responsible for trying offences including treason, murder, rape, arson and burglary; for punishing civil disobedience; for regulating the price and quality of food, ale, leather and footwear; and for early town planning. The offices of the Court Leet reflected these duties: the portreeves collected rents and revenues; the bailiffs summoned jurors and seized stolen goods; the aldermen policed the maintenance of public order in their wards; the constables enforced the law; the rhine ridders supervised the cleaning of gutters; and the shamble keepers, ale-tasters, searchers and sealers of leathers and of greenskins, and the cornhill keepers regulated the markets (Hedworth Whitty, 1934; Sheppard, 1909).

The official responsibilities of the Court Leet were eventually ceded to the Taunton Borough Council – although the loss of the town's charter between 1792 and 1877 prolonged its existence later than in most other towns. However, the historical link is tangential to its current purpose:

> That was the forerunner ... of local administration. But it's almost a play act, I mean, you tend to get fed up, we don't do anything except have an outing once a year. (Interview – I220 Court Leet Member).

> The Court Leet wasn't about history. They may pretended, that may have been the excuse for them to come together, but it was an excuse for social outings, dinners, of getting together, of the old established businesses in the town, of which there are of course fewer and fewer now, coming together and enjoying themselves. And, I suspect, influencing each other in terms of their opinions and their reactions to what was going on in the town. (Interview – I1347 Local Historian and Councillor).

The narrow base of the Court Leet is protected by its oligarchic invitation-only method of appointment. As representatives of established Taunton businesses, its members in 1995 included directors of the town's two major building companies, four solicitors, three chartered accountants, two chartered surveyors, a civil engineer, an insurance agent and the former manager of the County Hotel. It is perhaps indicative of the withdrawal of the business community from local government, that few members held official governmental positions; but the Court Leet is probably more important as the most exclusive of several inter-linked cogs through which the business community operates:

> You've got circles in Taunton, and there's a joke, or I find it slightly amusing, that if you've got the Masons, and I'm not a Mason, you've got a lot of people in there who were in the business community, and you put a ring around that, and there's Rotary... So in the Taunton Rotary, if we've got eighty members, which is quite large, I don't know, we might know more than one used to, it was probably fifty Masons, I never know, maybe out of ninety. Then we've got in Taunton a thing called the Court Leet. (Interview – I220 Court Leet Member).

The larger elite network, which spread out from the Court Leet through the Masons, the Rotary Club, the Chamber of Commerce, the Civic Society and the Soroptomists, linked in more active political actors, including both county and borough councillors, the Conservative leader on Taunton Deane Borough Council, directors of the health commission, NHS Trust and the training and enterprise council, magistrates and school governors (Figure 3.7). They still, however, represented a much lesser degree of potential influence than the same organizations were able to achieve during the mid twentieth century on the old Taunton Borough Council. As in many towns, the dominance of that council by businesspeople led to rumours about its internal practices:

> In the old days of the urban districts and rural districts, smaller councils, the accusation was very common at doorstep level, you know, 'it's just old Mr Jones and his mob, they run everything in this area, it doesn't matter what we do about it' you know, sort of thing. It really was accusations of a sort of inner mafia, in most cases aimed at rich and wealthy businesspeople, landowners, people like that ... It used to be quite prevalent in Taunton at one time because we had quite a few builders and estate agents and people in the real estate world who used to be well known leaders of the council, and there was always this accusation also that they had hands in the tills ... I mean, we've never had any proven cases here, but you know what the public are like. (Interview – I27 Taunton Deane Borough Councillor)

Central to such accusations was the role of the Freemasons. There are nine Masonic lodges meeting in Taunton, the oldest dating back to the 18th century. The foundation stone of the Taunton and Somerset Hospital was laid with a Masonic ceremony in 1810, and the centrality of the Masonic Hall (a former Catholic church) opposite the County Hall perhaps symbolizes the importance they enjoyed for much of the twentieth century. With over 200 local members during the 1940s, Masonic influence in the immediate post-war years expanded beyond local government into organizations such as the Taunton Operatic Society. The social and economic restructuring of Taunton in recent decades has blunted the influence of the Freemasons as with the rest of the business elite, and the questionnaire of councillors revealed only one declared Mason on Taunton Deane Borough Council in the mid 1990s. Nonetheless, the suspicion remains:

76   Contesting Rurality

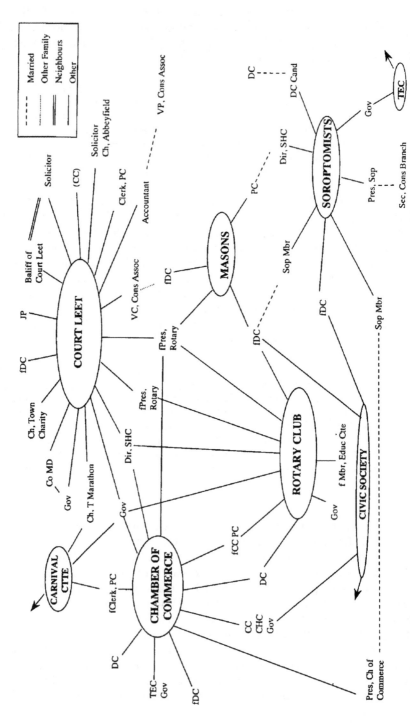

Figure 3.7 The Taunton business elite network

> As a chairman of planning you become very acutely aware to watch out for anything that perhaps smells anything to do with the Masons. I get totally paranoid at times about it, I mean, it is a subject that I have read quite a bit about and really it is just so horrendous. (Interview – I401 Planning Committee Chair).
>
> You must ask the question why? Why do they find it necessary to be in the Masons? And then you see the links into local businesses, building companies and things like this and you realize that there's a whole network there that you never become part of. (Interview – I6 County Councillor).

By the 1990s, the business elite may not have had the kind of direct, hegemonic, power that it once enjoyed on the old borough council; but that should not detract from the influence that the close-knit network of politicians and businesspeople could achieve, supported by the dominance of a pro-business discourse at national level. The process of local governance offers formal channels through which the representations of the Chamber of Trade and other organizations, such as the Soroptomists, can be directed:

> It is a pressure group and if it feels that there's something wrong it will say so, locally or nationally, and equally well it likes to take part in local consultation. So, my own club is doing something on the Somerset Structure Plan, I hope it will be responding to [the health authority's] purchasing intentions next year. It participates locally in a lot of things and really we're just giving a bit of our expertise back into the community through that. (Interview – I408 Former President, Soroptomists).

At the same time these organizations, along with the Rotary Club, the Civic Society, and more exclusively, the Freemasons and the Court Leet, provide elite spaces for discussion and decision out of which a 'business' approach can be developed for negotiating with local government over schemes such as the controversial re-development of Taunton town centre in 1996, and wider issues of housing and commercial development (see also Cloke *et al.*, 2000, on homelessness policy in Taunton). On topics such as these, the voice of the business elite cannot be ignored and their influence continues.

Story 4: The Party Set

The examples discussed in this chapter have so far concentrated mainly on social networks between people that exist independently to their political activity, but nonetheless may have an influence on it. However, for many people, involvement in local politics leads to new contacts, new friendships and new social networks. When it comes to patronage or exercising influence, contact through these political networks may be more important than other social relations. For the more established elites the distinction between political and social networks is blurred. Social activities are important to the organization of the Conservative Party and a number of councillors were asked to stand having become known through involvement with the local Conservative Association; but official party events are not the only forum in which party business is conducted:

> The whole countryside is shot through with these threads of Toryism and Conservatism which have their origins in ages past, and although, for example, the Royal National Lifeboat Institution to take but one, although that's not overtly a Tory front organization, it is composed basically, certainly in the organizational levels, of people who adhere to Tory principles. And they get together, it doesn't matter what organization it is, they get together and they talk Toryism, and that's the kind of organization which bolsters the idea of Conservatism and Toryism throughout the country. (Interview – 121 Taunton Deane Borough Councillor).

In contrast, Labour and Liberal Democrat activists encountered the problem that the identification of established social organizations with Conservatism was so entrenched that either they found themselves excluded completely, or at least found that such networks were not open for furthering their own party political activities. Instead alternative political networks had to be developed. To some extent the social and voluntary organizations within the 'emerging liberal elite network' perform this role, but it is notable that most of the organizations in that network have an implicit political agenda (such as Friends of the Earth or the Taunton Association for the Homeless), whilst many individuals involved with the more established voluntary organizations included (such as the Citizen Advice Bureau or the Council for Voluntary Services) became involved as representatives of the council. There is not the same emphasis on social activities as in the other networks:

> A typical group of Liberal Democrats would be far more inclined to go for say a chairman rather than the Mayor. We're very good at sort of wearing sack cloth and ashes and not getting involved in social events because they might be the cost of a couple of hundred pounds for a buffet and things like this, and so on. In other words, we do try and detach ourselves, there is a tendency of some people to detach ourselves away from the civic life of the town. (Interview – 16 County Councillor).

Consequently, the social network of leading Liberal Democrat actors is mediated not through the social activities of established organizations such as the Court Leet or the Lieutenancy Dining Club, but rather through social activities organized by the Liberal Democrat party:

> A lot of our social life revolves around friends within the party. We go to a barbecue for fundraising, we're going to a barbecue on Saturday night to raise funds for the party, and it's good fun. We have a dinner up at the village hall every year in October, and a speaker and raise funds, which promotes a social life for political reasons. (Interview – 121 Taunton Deane Borough Councillor).

The promotion of 'a social life for political reasons' performs a useful function in connecting socializing with fundraising, but equally importantly it creates elite spaces that are explicitly identified with the Liberal Democrats and within which issues and strategies can be discussed and contacts formed which might be drawn upon through recruitment and patronage.

Perceptions of Elite Networks

The existence and use of elite networks therefore permeates all parts of local governance in western Somerset and all political parties. Indeed, there is widespread consensus that, as one Labour activist expressed it 'networking is very very very important'. Where opinions diverge is in the perceptions of how elite networks are used, how open or closed they are, and their contribution to the wider political system. It is perhaps the role of the long-standing elite networks associated with the traditional rural elite and the business community that provokes the greatest disagreement. For many of the political actors who are not included in them, they are perceived with suspicion, viewed almost as cartels of hidden power, a perception relayed by the emphatic use of terms such as 'mafia', 'covert influence' and 'establishment':

> There are covert influences in West Somerset which are hard to pin down but their existence over many years is a cause of concern. No idea of full make up of the group, who leads it or how they function. Probably outside the council. (Questionnaire – Q73 West Somerset District Councillor).

> The local 'mafia' who operate through political, commercial, economic and social power [are the most influential]. While traditionally headed by the local 'wealthy' families this situation is evolving to bring into the network of contacts anyone who is perceived as having the potential to influence local affairs. In return new recruits must show loyalty to those at the top of the system and be happy to feel they belong and receive their share of the spoils from being on the winning side. (Questionnaire – Q265 West Somerset District Councillor).

The picture created by such comments is one that mystifies the role of the 'establishment'. The alleged secrecy that prevents observers from seeing how the elite's mechanism works – and thus allows speculation about a mafia-type organization – is an essential part of building up the image of a covert power source. Equally important is the idea that membership of the 'establishment' is uncertain and restrictive. Although the second quote above talks about the 'establishment' expanding from its traditional base, it also contends that the 'establishment' is only recruiting those who fit its discourses and are prepared to accept its hierarchy – a sentiment echoed in the frustration of a senior councillor at not being able to get into the network: 'We're not invited. We're not invited.'

Others, however, dismissed the idea of the 'establishment', arguing that social, economic and political restructuring has produced a more open power structure:

> If you are thinking in terms of Establishment v. non-Establishment, that was over 30 years ago. Taunton is no longer dominated by Rotary etc. Too many chain stores dominate the scene. The Church is too busy playing church games to be relevant. (Questionnaire – Q24 Taunton Deane Borough Councillor).

Among some of the political actors who were linked into the 'traditional rural elite network', the suggestion that they were part of the 'establishment' was accepted with resignation, not considering themselves as being 'establishment', but conceding that others might place them as such. One former county councillor sought to define the establishment differently, as being a term which was only relevant with respect to the landowning or commercial families such as the Luttrells and the Clarks – a definition which both excluded himself from the establishment and removed some of the malice linked to the term. Another former county councillor, accepting that he was himself part of the 'establishment', suggested that its function was constructive:

> It's all part of the hand-in-hand structure of the unwritten constitution of this country, which is full of the checks and balances, isn't it, you know. Hence, be very careful if one is taking any radical move because if you knock a brick out of a wall it will loosen the others. (Interview – 1190 Former County Councillor).

From this perspective, the 'establishment' is not a covert power elite that pro-actively seeks to influence local governance, but is purely a reactive defence against radical action by elected councillors. Interestingly, this perception of the establishment's function is not wildly different from that expressed by some of its opponents, albeit with rather different connotations:

> You become painfully aware that whatever successes you have at the ballot box, the 'establishment', and I use that term quite deliberately and loadedly, is there. It's in evidence. It's going on and it's 'this lot is only passing by, we're here', and they are really quite entrenched. (Interview – 129 Taunton Deane Borough Councillor).

However, whilst many Liberal Democrat and Labour activists view the networks of the traditional 'establishment' as a negative force blocking the ambitions of elected government, they perceive their own networks quite differently. Whereas the 'closed' networks of the old elite are seen as being about restricting the dispersal of power, they argue that their own 'open' networks are about diffusing influence:

> It's open networks because you're always wanting to look for fresh talent, fresh faces, fresh fundraisers to bring in, whereas some others tend to be, well, if you don't know who they are this is where one's suspicions rise, when you don't know who they are. (Interview – 129 Taunton Deane Borough Councillor).

For this group of people, their social and professional networks are a form of 'outreach', consulting people outside the formal decision-making process and gathering knowledge from people involved in a wide range of activities:

> I think networking in the widest possible sense is vital for the cross-fertilization of ideas. The informal side of council work is just as important in outreach as it is in the formal decision-making process. Communication and the understanding of ideas, I place a great deal of emphasis on that. (Interview – 1400 Council Leader).

If you do have outside contacts it is easier really, because when something comes up you can ring up friends and say, well, you know, 'I've got this, what do you think?' 'What would...?' etc., and if you represent an area too, you have contacts in that area. I represented two villages and there were a lot of people that I knew who were in sort of different stratas of life, who one could, sort of, use as sounding boards. (Interview – 1120 Former Taunton Deane Borough Councillor).

People seem to overlap there interests, and certainly in terms of the community health council, we find that we get to hear about all sorts of things from people involved in lots of different things. It's not a political thing. And I think the same applies to the politics of the area. (Interview – 1403 Chair, Community Health Council).

In drawing a distinction between the constructive, open, use of their own networks and the negative, closed nature of their political opponents' networks, liberal politicians invite the question of where one becomes the other. This problem is recognized by some:

There's a danger that you do establish what I call, what I suppose is the old boy network. I'd like to think that if you are scrupulous in the way you go about things, you do not mix the two. You use your social contacts, the informal network, for the gathering of information, for the consultative aspect of this, without actually in any way prejudicing the formal means, your process for establishing your priorities. I like to think that is true. I suspect that there will be always those who will see somekind of conspiratorial things if they so want, but I'd like to think ... I'm afraid that's a fact you've got to live with, but I'd like to think we are open in that sense, when it actually comes to our proposals which go before the council. (Interview – 142 Taunton Deane Borough Councillor).

The repetition of 'I'd like to think' in the above quote is telling phrase. Whatever the practice of elite networks the perception of their composition and activity will be contested by individuals informed by different discourses.

Conclusion: Elite Networks and the Rural Local State

The case study of Somerset in the mid 1990s presented in this chapter demonstrates that elite networks continue to be a feature of the contemporary rural power structure, and that such networks are employed in the recruitment of political activists, the appointment of members of local governance bodies, and the exercise of influence in local governance. However, simply identifying such networks and describing accounts of their operation does not show that rural local governance is elitist in its policy implementation. Indeed, such a connection is more difficult to prove and perhaps only becomes evident in cases of the most blatant use of personal contacts to achieve favoured outcomes. For example, in 1971, Williton Rural District Council was accused by a local parish council of 'underhandedly' allocating public housing. As the accuser stated at the time, 'if you are not in favour with a councillor, your chances of obtaining a house are slim.'(*West Somerset Free Press*, 20 February 1971).

The greater openness of local government and the creation of tighter safeguards against malpractice probably mean that such overt abuse of elite networks is uncommon today. Rather, networks are used to exercise influence in a much more subtle manner. Thus, as one councillor described earlier in this chapter, knowing an officer socially outside the context of the council may help in persuading that officer to prioritize a particular activity which might otherwise have sat in a pile of other tasks waiting to be done. This is not so much an abuse of personal contacts, but rather the employment of social capital in the same way that another councillor or activist might use their personal capital of communication skills, or education, or knowledge of the system to a similar effect. Furthermore, this 'quiet' exercise of influence is indicative of a much more complex structure of power relations than envisaged in a model of resource-derived causative power.

Additionally, we should be careful not to over-state the scale of elite networks in local governance. They are not hegemonic cabals to which all those in local governance positions must belong, nor is all local political business conducted behind their closed doors, as one councillor stressed:

> If we take the local magistrates, I have never ever since I first came to Minehead, even since I joined the council, I don't think I've ever bumped into one, had a drink with one, been invited to somebody's house for dinner, whatever. The local council - almost the same. (Interview – 178 West Somerset District Councillor).

This does not prevent suspicions forming about the amount of power which elites have and how they practice it. Just as significant as the actual organization and activity of elite networks are the perceptions of them. In other words, the mechanics of the local power structure, as represented by the elite networks, cannot be separated from discourses of locality and power, through which they are attributed with power.

So what difference has the existence of the particular elite networks described in this chapter made in shaping the local polity in western Somerset? The answer is that elite networks have acted as 'gate-keeping mechanisms', protecting a particular local political culture. The use of elite networks for recruitment and patronage helped to ensure that councillors and other local governance office holders tended to have similar ideological outlooks and similar ideas about the character of the locality, thus creating an impression of consensus. Meanwhile, close ties with certain organizations – whose officers were members of the elite network and whose meetings formed 'elite spaces' – served to prioritize particular interests.

However, the effect of elite networks is perhaps more apparent in what is *not* done, than in what is. For example, it is unlikely that a council leadership integrated into the business-centred elite network would have supported the redevelopment of Taunton town centre embarked on by Taunton Deane Borough Council in 1995, which was opposed by many prominent businesspeople who feared that the proposed pedestrianization and traffic calming schemes would cause disruption, reduce access, increase congestion and discourage people from shopping in the town centre. Similarly, an attempt by Somerset County Council to ban staghunting on its land mounted in 1993 could not have occurred earlier as previous council leaderships had always been linked to the hunts through the 'traditional rural elite network' (see Woods, 1998a).

As such, elite networks have contributed to structuring a local state which is distinctively *rural*. The use of social networks for recruitment and patronage and of exclusive elite spaces for lobbying and informal decision-making helped to extend the dominance of the landed establishment over local governance – originally founded on feudalism – well into the early twentieth century, and by extension, guaranteed that local governance was dominated by people whose view of rurality was influenced by historic discourses of stewardship and the 'country gentleman' and which prioritized the interests of agriculture and field sports. Alternative ideas – such as radical environmentalism – or interests – such as those of the rural working class – which may have resulted in different policy priorities were excluded because their proponents were excluded from the elite network, and hence did not have the same direct contacts with policy makers, or ability to construct networks of influence, or access to elite spaces. Even with the fragmentation of the power structure in the late twentieth century and the emergence of a new elite network of middle class actors broadly associated with the Liberal Democrat and Labour parties, which has resulted in elected local government adopting policies reflecting the interests of middle class incomers, the majority of the local population are not connected into elite networks, do not have the direct contacts with decision-makers, do not have access to the elites' backstages, and are not able to promote their interests and their ideas of rurality and locality with the same effectiveness.

Chapter 4

National Politics and Rural Representation

Introduction

The previous two chapters have focused on the distribution of power at a local level in rural Britain. However, the scope for the exercise of local power is limited. The competences of local government are determined by the national state, and the autonomy of local authorities to act as they see fit was significantly restrained as part of the restructuring of the state in the late 20th century. Even before the powers of local government were reduced, local leaders in rural society had to respond to national policies and political events than impacted on the countryside, and many key aspects of rural government were controlled from the centre, not through the local state. As such, the governance of rural space is regulated primarily through national politics and thus elite groups within rural society must ensure that their interests are appropriately represented in national scale politics.

Historically, rural representation depended on the personal influence of individuals who transcended the different scales of the state. The landowner who was simultaneously chairman of the parish council, a county councillor or alderman, a magistrate, and a Member of Parliament, or a member of the House of Lords, was able to make sure that their interests were represented at all levels. Others relied on the social contacts they had with members of the national elite through family, school ties or London clubs. As long as the national political elite was primarily drawn from the aristocracy and landed gentry, rural interests were represented at the heart of government and there was no need for an explicit rural lobby to be active.

From the 1920s, however, as hegemonic power ebbed away from the landed elite at both national and local levels, a division of labour became necessary. Whilst the agrarian and small town business elites came to prominence in rural local government (see chapter two), the responsibility of representing rural interests at a national scale was assumed by a triangular alliance of the National Farmers Union, the Country Landowners Association and the Conservative party (the Council for the Preservation of Rural England which was founded in the same era played a different role and is discussed further in chapter seven).

The glue that held together the various parts of this division of political labour was Conservatism. Although it has been argued that Conservatism should not be regarded as an ideology or a doctrine, for much of the twentieth century it could be described as a coalescence of political beliefs and positions that was founded on the protection of property rights, the maintenance of social order, and the balancing of tradition and progress. As chapter two described, these features were the core elements of the dominant discourse of power in rural society during the mid part of the century, but they also found expression in the policies pursued by the Conservative party at a national level.

This chapter explores the characteristics of rural Conservatism in more detail, before examining the contradictions within Conservatism that eventually led to the weakening of rural representation through the party and the subsequent need to construct an explicitly rural lobby.

The Rise of Rural Conservatism

The collapse of the Liberal party after the First World War left the Conservatives with a virtual electoral monopoly in most of rural Britain. Although the Liberals retained a hold in much of rural Wales and northern Scotland, as well as in a few dispersed parts of England, the Conservatives moved into a clear majority position in most rural constituencies during the 1920s and 1930s. As the alternative was the strengthening Labour party, with its urban base, its allegiance with the trade unions, and its commitment to nationalization and class equality, the rallying of the new agrarian elite behind the Conservative party was a rational, self-interested, move. What was remarkable, however, was the speed and thoroughness with which the NFU and CLA were able to integrate themselves into positions of influence in the party.

At a local level, branches of the NFU exerted considerable influence over the selection of Conservative parliamentary candidates, pursuing a strategy of getting farmers into parliament. When in 1922, the chairman of the Taunton NFU branch, Robert Bruford, was selected as the Conservative candidate for Wells (and became one of the first cohort of tenant farmers to be elected to parliament) (see chapter two), he had first been selected as the candidate of the local NFU branch. Similarly, in 1921 the Herefordshire North Conservative Association had asked the local NFU branch to select their candidate for them (Moore, 1991). In spite of such arrangements, however, the number of farmers and landowners in the parliamentary Conservative party during the inter-war period was relatively small. Only an eighth of the parliamentary party in 1924 were landowners, and over a third of the Conservatives MPs elected for rural constituencies in 1923 had never worked in agriculture (Moore, 1991).

The influence of agricultural interests in the Conservative party therefore rested on active lobbying by the NFU and CLA. The NFU, which was still to some extent an outsider group, gained access by sponsoring MPs and organizing sympathetic MPs into a parliamentary committee. The CLA was in a more privileged position due to the nature of its membership. Sir Robert Sanders and the Earl of Selborne, both Agriculture Ministers during the 1920s, were members of the CLA, as was Sir George Courthope, chairman of the Conservative agriculture committee. In 1924, four members of the CLA executive committee were appointed to ministerial office, including Edward Wood who became Minister for

Agriculture (Moore, 1991). Indeed, Moore observes that it was sometimes difficult to ascertain exactly on whose behalf some politicians were acting:

> An MP could join an NFU or CLA delegation to the Ministry of Agriculture on one day, and speak at the Conservative Party conference the next. Loyalties to countryside and to party were so closely enmeshed that it would be impossible and inappropriate to separate them. (Moore, 1991, p. 344).

Political opponents including David Lloyd George and Ramsay MacDonald accused the Conservatives of pampering agriculture because of the influence of small interest groups tied to the Tory party, and suggested that agricultural policy was mis-managed as a result (Moore, 1991). The Conservatives, in contrast, openly celebrated their attachment to agriculture, which they considered to be a product of a traditional affinity with landownership, the family and occupational backgrounds of MPs, and a genuine empathy with the vulnerable economic position of farmers. Either way, ruralism became a core element in the 'new Conservatism' developed during the mid 20th century, as illustrated by the emphasis placed on property rights, the notion of the organic society and on national identity as reflected in the English rural landscape and culture.

Property Rights and the Organic Society

The clearest correspondence between Conservative ideology and the discourses of rurality reproduced by the local rural elites of the mid twentieth century is found in the Conservative commitment to property rights. As Barnes (1994) comments, the virtue of property ownership was to many Conservatives essentially axiomatic, such that the centrality of property to Conservatism 'is such that Conservatives assume that it needs little if any justification' (p. 325). A justification was provided, however, by Quintin Hogg (later Lord Hailsham) in one of the key texts of the 'new Conservatism':

> Conservatives believe that the incentive to possess property by legitimate means is one of the most valuable aids to the production and increase of wealth. Provided that others are not thereby impoverished or harmed they consider that the possession of large fortunes is a good – a good both absolutely and relatively because it tends to the diffusion of economic power and away from its concentration in the hands of the Government. Nor have Conservatives the smallest objection to the existence of a leisured class which uses its leisure well. On the contrary, however short the working hours of life become, Conservatives believe that a leisured class has much to bring to society – both in culture and wisdom. (Quintin Hogg, *The Case for Conservatism* (1947), quoted in White, 1950, pp. 93-4).

Hogg's statement of the Conservative belief in property is not the assertion mobilized by later Conservative politicians that all citizens have a right to property ownership (manifest, for example, in the Thatcherite policy of council house sales). Rather, Hogg explicitly makes the assumption that property ownership is unequal and suggests that the existence of a 'leisured', property-owning class is a good thing. Hence, Hogg's Conservatism also respected and accepted social hierarchies as part of a structure of 'traditional authority' that operated within an 'organic society'. This belief, in particular, distinguished Conservatism from the radical promises of Socialism, warning that change must be gradual and natural:

> Conservatives ... believe that a living society can only change healthily when it changes naturally – that is, in accordance with its acquired and inherited character, and at a given rate ... Moreover, Conservatives also draw the moral that there is an advantage, even from the point of view of those desiring radical change, in preserving the mystique of traditional authority. (Quintin Hogg, *The Case for Conservatism*, quoted in O'Gorman, 1986, p. 68).

Indeed, for Barnes (1994), 'Conservatism arises directly from the sense of a continuing social order, and this presupposes the conception of society as organic, with individuals valued not simply in themselves but because they are rooted in a particular time and place and order' (p. 321). The Conservative party could hence be positioned as the 'natural' party of the countryside, as imagined as an organic community, and played on nostalgia and fear of change in securing its rural vote.

One legacy of this discourse is that rural Conservatism has persistently been suspicious of people and lifestyles that do not conform to community norms. This has, at various times and in various places, been manifest in racism, homophobia, extreme localism and other forms of discrimination. Particular intolerance has been expressed towards 'outsiders' who are perceived to disrupt or destabilize property rights, as has been evident in the attitude of some Conservative politicians towards 'new age travellers' at both national and local scales. Nationally, it was a Conservative government that introduced legislation to attempt to restrict the mobility of travellers in the Criminal Justice and Public Order Act of 1994 (Halfacree, 1996; Sibley, 1997). As Halfacree (1996) records, Conservative parliamentarians speaking on the legislation evoked this traditional strand of Conservatism in emphasizing the perceived threat posed by travellers to property rights and the order of the countryside:

> New age travellers appear to have no wish to establish themselves or reside on authorized sites, but simply want to roam through the countryside unchecked. (Tony Baldry MP, quoted in Halfacree, 1996, p. 58).

> These people camp in an area and then chase the sheep, rob and do other things. (Lord Stanley, quoted in Halfacree, 1996, p. 59).

> The new age travellers displayed some dreadful antics: they invaded peaceful countryside, decimated peaceful villages, went on the rampage and had raves lasting two or three days, showing a total disregard for the area. (Nigel Evans MP, quoted in Halfacree, 1996, p. 62).

The Criminal Justice and Public Order Act gave police and local authorities powers to remove travellers who they believed to be illegally camped, to seize vehicles and to pre-empt the movement of travellers by stopping vehicles believed to be heading for a 'trespassory assembly'. In short, the Act sought to impose a spatial order on travellers in which they would only be accepted as legitimate occupants of rural space if they were resident on official authorized sites. An earlier piece of legislation, the Caravan Sites (Amendment) Act of 1993, had placed a statutory requirement on local authorities to provide residential or transit sites for new age travellers. However, the designation of such sites has become a contentious issue in the local politics of many rural areas, with campaigns frequently led by Conservative councillors, mobilizing the same ideas the threat to the organic community. In Somerset, for example, opposition to the provision of statutory facilities for travellers was a key plank of the Conservative campaign in the 1995 local elections, with one election leaflet stating that:

> YOU CAN SAVE THE DEANE FROM LIB DEM PLANS TO BUILD NEW AGE TRAVELLERS' CAMPS. Recently the Lib Dems announced a new initiative – build 'New Age Travellers Camps' all around Somerset. Residents in both rural and town areas have already had nasty experiences when such people move into a locality. How can the Lib Dems consider this a priority at such a time (The Middlezoy camp cost something around £250,000). We need YOUR PROTEST VOTE to fight this. (Conservative election leaflet, Somerset, May 1995).

The same themes were reproduced by a Conservative councillor on Somerset County Council in a debate on the provision of emergency stopping places for travellers in May 1996, as reported in the local press:

> 'The big difference between them and the 500,000 we're responsible to is that we pay our way – they don't', said Mr Roach. He claimed that Somerset would become 'the soft touch for these people' and neighbouring counties and EC countries would see 'what a mess we're making'. He accused the travellers of playing on the heart strings by taking children to the meeting and said, 'These people aren't from Somerset' ... In a motion, Mr Roach hit out at the expense of setting up emergency sites, detrimental effects on tourism and nuisance to landowners. (*Somerset County Gazette*, 17 May 1996).

Thus the argument against the provision of travellers' sites in Somerset drew not only on issues of financial efficiency, but also on ideas of localism and the perceived threat from alternative lifestyles on received notions of rurality. In this way the discourse of the organic community and the ideology of property rights continue to inform the practice of Conservative politics in rural areas.

National Identity, Rurality and Conservatism

The third key component in the Conservative discourse of rurality was the conflation of Conservatism, ruralism and British, or at least English, national identity. The association of rurality and national identity has ancient origins. It can be identified in classical literature and the poetry of mediaeval England, and is also prevalent in many national cultures (Short, 1991; Williams, 1973; Woods, 2005a). Rural areas are venerated as national heartlands in part because of the importance of territory to national identity, and consequently the identification of rural landscapes as symbols of national spirit and character; in part because of the perceived purity and stability of rural life and the persistence of traditional values and culture; and in part because of a moral geography that represents the city as having been corrupted by foreign influences (Daniels, 1993; Ramet, 1996; Short, 1991). In many countries, this discursive framework led to a celebration of the rural peasant class such that the identification of national identity and rurality could have progressive overtones. In England, however, the progressive movement tended to promote urban, industrial, society – and the labour of the working class that produced it – as the expression of national identity, whilst the positioning of the countryside as the symbol of Englishness was driven by the aristocratic elite. Crucial to this development was role played by members of the landed classes in building the British Empire and their memories of the great estates of the English countryside as the homeland they had left behind. As Daniels (1993) commented,

> The very global reach of English imperialism, into alien lands, was accompanied by a countervailing sentiment for cosy home scenery, for thatched cottages and gardens in pastoral countryside. Inside Great Britain lurked Little England. (p. 6).

As the empire began to be dismantled in the mid twentieth century, the countryside was the space to which erstwhile colonial administrators and settlers returned, often purchasing old manor houses vacated by the shrinking squirearchy or building large new houses in open countryside that reflected styles of colonial architecture. However, the countryside to which they returned was not the rural England of their colonial imagination. It was the rural society that was undergoing the processes of social, economic and political restructuring and as such became re-positioned as a vulnerable, endangered place. The battle to conserve the true 'England' was thus fought on two fronts – the attempt to hold on to a colonial lifestyle as the Empire contracted, and the defence of the English rural idyll at home:

> The symbolic significance of the homeland steadily changed, for increasingly this now became 'home', marking a shift to a more inward-looking, introspective, domestic, form of rurality (most clearly in evidence in the 'Home Counties'). And as the rural vision became introspective, its essential features came to be seen as both timeless and fragile – an endangered 'deep England'. (Lowe *et al.*, 1995).

In this way several different imperatives – the interest of the landed gentry in retaining their status and influence; the protection of the agricultural economy from urban influence; the maintenance of the rural 'organic society'; and the defence of the nationalistic vision of rural England – all became aligned in Conservative ideology. The significance of the countryside to the Conservative vision of England was most famously articulated by the then Prime Minister, Stanley Baldwin, in a speech delivered to the Royal Society of St George in 1924:

> To me, England is the country, and the country is England ... The sounds of England, the tinkle of the hammer on the anvil in the country smithy, the corncrake

on a dewy morning, the sound of the scythe against the whetstone, and the sight of a plough team coming over the brow of a hill, the sight that has been seen in England since England was a land, and may be seen in England long after the Empire has perished and every works in England has ceased to function, for centuries the one eternal sight of England. (Stanley Baldwin, quoted in Paxman, 1998, p. 143).

Yet, as Paxman observes, even by 1924 the picture of rural England described by Baldwin was pure fantasy:

There was absolutely nothing eternal about any of these sights or sounds. The scythe was already being replaced by harvesting machines, and as the internal combustion engine moved in, the blacksmith was reduced to making shoes for the ponies of the children of the businessmen who were buying up the cottages of farmworkers driven from the land. By the time of this speech, England had been a predominantly urban society for seventy years. The vast majority could no more have recognized the rerrk-rerrk call of a corncrake than they could have parsed Sanskrit. (Paxman, 1998, p. 143).

Nevertheless, this discourse helped to give rural interests a privileged position in the Conservative party at a national level and contributed to the construction of the Conservative hegemony in the countryside. Moreover, explicit displays of patriotism have remained an important part of Conservative party culture in rural areas to the present day. When the Watchet branch of the Conservative party in Somerset celebrated its twenty-fifth anniversary in 1995, the event's finale reportedly featured the branch president 'parading a large Union flag into the hall followed by members of the committee to *Rule Britannia* followed by *Jerusalem* and *Land of Hope and Glory*' (*Somerset County Gazette*, 29 September 1995).

Even as the Conservative electoral hegemony in rural areas was broken in the 1990s, the discursive association of rurality and British national identity was employed by new parties that sought to attract dissident Conservatives. Relatively high levels of support for the United Kingdom Independence Party (UKIP) in rural constituencies are in part founded on this association and the UKIP has distributed leaflets at the countryside marches in London. A similar attempt to exploit rural dissatisfaction has been made by the far-right British National Party (BNP) which since 1997 has periodically produced a 'British Countryman' newspaper aimed at rural voters. The contents of the newspaper have an unsettling resonance with some of the key ideas behind the association of rurality and national identity – notably that rural society represents a purer, more true, British way of life; that threats to traditional rural activities are also threats to Britishness; and that the threat to the countryside comes from an urban society that has been corrupted by foreign ideas.

Arguably, traces of the discourse can also be found in instances of both explicit and implicit racism in rural areas – now acknowledged to be a prevalent but under-reported problem (Agyeman and Spooner, 1997; Woods, 2005a) – as well as in local protests against the siting of detention centres for asylum seekers in rural locations (Hubbard, 2005).

The Contradictions of Rural Conservatism

The Conservative party has always prided itself on being a broad church, yet its breadth has also been part of its downfall. The discourses of the organic society, of property rights and of an English nationalism rooted in the countryside may have achieved prominence within the party during the twentieth century, but it has always been one strain among many in modern Conservatism. The inherent preservationist instinct of rural Conservatism, for example, has long been in tension with the generally pro-development bias of the party's economic policies. Moreover, the emergence of a more radical Conservatism from the 1970s onwards promoted a new approach to social and economic issues that was frequently directly contradictory to key tenets of the traditional rural Conservative discourse.

Conservative activists in rural areas will often identify changes in societal attitudes as one of the reasons for the decline in the party's support. One branch chair in Somerset, for example, blamed the 'antics of the royals' and the liberalism of the Church of England for his party's problems, whilst a long-serving Conservative councillor argued that the party had become out of tune with the nature of contemporary society:

> I remember going to conferences when I was a Young Conservative and you only had to get up and say 'The Conservative party stands for God, Country and Empire', and you'd get an ovation ... But that was the culture from which it came. Society then had parameters. This is what the problem is today, there are no parameters, you have the young culture, but the word 'culture' means that there are parameters outside which people who are in that culture, you know, will not overstep ... And then you had a strong religious mood, sixty per cent of people went to church, or would be considered church-goers even if it wasn't a Sunday ... You had an immobile population because the only thing they had was a bike ... and you had a non-anonymous population, the shopkeeper knew me, parents of friends knew me, the vicar knew me, the teachers knew me, a whole load of people knew me, so I behaved ... So, the Conservative party over all these recent decades has progressed from that base, and that base has diluted over the decades, because there's television, radio, video, all sorts of things have released people from the need to be within the parameters. So therefore, the strength of the Conservative party was the fairly rigid climate, class-ridden yes ... [Now] we have a highly mobile population, a highly secular population, a population that has been affected by your normal human nature, greed of acquisition, of a house, wall to wall carpets, a car etcetera ... The nature of community has changed. Now this trend again militates against the party which actually conceived the boundaries ... So, QED, the Conservative party values, strictures, history, are now not the avant garde of the new thinking, the new community. (Interview – 150 Borough Councillor).

Significantly, the trends that are identified here – individualism, materialism, consumerism, the weakening of class identity and of residential communities – as well as the credit boom that made many of these changes possible, are all most strongly associated with the policies of Thatcherism pursued by the Conservative government from 1979. Thatcher's statement that 'there is no such thing as society' in a 1987 magazine interview was a summary dismissal of a core part of the rural Conservative heritage.

Even within rural areas the values of the traditional Conservative elites came under challenge from new activists, exaggerating tensions in emphasis that already existed between agrarian and mercantile Conservatives. These tensions could be accommodated whilst the two elites exercised their separate authority over rural and small-town councils, but the 1974 reorganization of local government threw the two groups together as allies on new county and district councils. Young (1994) notes that every major reform to local government up to the end of the twentieth century had been introduced by a Conservative government; but whilst the 1889 reforms that created county councils had vindicated '[Prime Minister] Salisbury's judgement that landed power would be consolidated rather than dissolved' (p. 406), the 1974 reorganization is described by Young as 'a politically expensive mistake' (p. 435) for the party:

> Its impact on county council representation was to realize, almost a century late, Joseph Parke's vision of a county elite 'cut down by the scythe of reform'. The county notables, some of whose families had dominated their county councils since 1889, virtually disappeared. The close ties between social leadership and Conservative politics were fractured, and county politics would not be the same again. (Young, 1994, p. 435).

Instead, the initiative passed to a 'new breed' of more politically motivated party activists and councillors. The transition has been described by some commentators as the usurping of traditional 'rural' Tories by 'Town Tories' such as accountants and lawyers (Cloke, 1990; Newby *et al.*, 1978). However, whilst the concept of 'Town Tories' has a relevance in counties such as Suffolk where reorganization united non-partisan rural authorities with highly politicized urban authorities, it is more problematic in counties such as Somerset, where both rural and urban councils had been non-partisan. Rather it can be argued that the decision by Conservative Central Office to seize the opportunity of the 1974 reorganization to politicize rural local government changed both the nature of election campaigns and the qualities looked for in candidates, thus encouraging a new cohort of councillors to emerge in both urban and rural wards, as one Somerset councillor remembers:

> There were a lot of new faces and they were rather different Conservatives. I'm not saying that they were unpleasant Conservatives, I mean, I count many of them as my friends now, but they regarded themselves as being there to pursue the policies of the Conservative party, and they kept a very wary eye on Whitehall and the thing changed. (Interview – 1156 Former county councillor).

Nationally, the weakening of agrarian influence in the Conservative party was less sudden, occurring steadily over a number of decades. The proportion of Conservative Members of Parliament drawn from farming backgrounds declined from around one tenth in the 1950s to just ten out of 336 in 1992 (Criddle, 1994). As Criddle observes, the changing occupational background of MPs 'mirrored, in the retreat of Bar, army, and land, the switch of electoral power from the shires to the suburbs: from estate owners to estate agents' (p. 161). The influence of the landed elite lingered longer, however, in the more senior reaches of the party. The governments of Macmillan and Douglas-Home in the 1950s and 1960s were notorious for their aristocratic flavour, even though the parliamentary party was already dominated by the professional classes. Margaret Thatcher's first cabinet in 1979 still included seven members of the aristocracy or landed gentry, although only two cabinet members – William Whitelaw and James Prior – had had any professional involvement with agriculture. By her third election victory in 1987, the number of aristocrats in the cabinet was down to four, and only Whitelaw and John Wakeham farmed.

As agricultural interests became less important to the Conservative leadership, the relationship between the party and the farming unions were increasingly strained under the Thatcher government. The political alliance between the unions and the Conservative party, and the unions' involvement in a close-knit agricultural policy community, had succeeded in guaranteeing substantial subsidies for farmers that were subjected to relatively little political scrutiny. It obviously remained in the farmers' interests for subsidies to continue and for the structure of agricultural policy to be left unchanged. Yet, state support for the agricultural industry was directly antithetical to the logic of Thatcherite economic policy, and the expenditure of 70 per cent of the European Union's entire budget on the Common Agricultural Policy (CAP) jarred with the government's cost-cutting battles with Brussels. Thus, whilst the Thatcher government never mounted a serious assault on farm subsides for fear of the electoral consequences, it did support and implement minor reforms to the CAP such as the introduction of milk quotas, in the face of opposition from farmers, and licence was given to maverick MPs like Richard Body to advocate a radical dismantling of the subsidy system (Body, 1982).

The post-Thatcherite Conservative party has continued to enjoy a mixed relationship with the countryside. The handling of the BSE crisis by the Major government in 1996 (see chapter six) marked a decisive break between the Conservative party and the farming lobby and provoked the defection of some farmers from the Conservatives, contributing to significant swings away from the party in several of the worst-affected rural constituencies in the 1997 election. At the same time, the loss of urban Conservative seats in the 1992 and 1997 elections means that the parliamentary Conservative party is proportionally more rural today than at any time in the last century. Since 1997, the party has also made determined efforts to associate itself with the emergent rural lobby, particularly during the leadership of William Hague, who as MP for Richmond represented one of England's most rural constituencies.

However, the reassertion of a rural Conservatism has been expedient rather than ideological and different visions of the countryside are articulated by different wings of the party. Communitarian Conservatives have re-captured some of the components of the traditional discourse by arguing for the need to revive and protect rural communities, advocating policies of decentralization and deregulation accompanied by support for voluntarism – an agenda laid out by the future

Conservative front-bencher, Damian Green, in a Social Market Foundation pamphlet (Green, 1996) and also reflected in much of the content of the Major government's Rural White Papers. At the same time, a more libertarian approach has been reproduced by groups such as the Social Affairs Unit in portraying the traditional rural community as an 'oppressed minority' and in defending hunting, opposing the 'right to roam', and advocating support for road-building, the development of greenbelt land and the deregulation of agriculture (Mosbacher and Anderson, 1999).

Hence, whilst in electoral terms the Conservative party continues to be the leading party in rural England (although not rural Scotland or rural Wales), its support is based on a coalition of a range of different interests and ideological positions and the party can no longer be identified as the primary means of rural representation at a national level.

The New Electoral Geography of the Countryside

The rural-urban cleavage in British electoral geography reached its apex in the 1970s. During the first half of the twentieth century the general pattern of Conservative representation was broken by pockets of Liberal and Labour support and by the occasional Independent MP. This complex geography reflected the significance of local factors in elections, including patterns of religious affiliation and economic structure as well as the following attracted by candidates as individuals. From the 1960s, however, there occurred a process, described as 'the nationalization of rural politics' by Johnson (1972), in which class allegiance asserted itself as the primary factor in shaping party support in both urban and rural areas. The more middle class complexion of the rural population favoured the Conservative party, and this tendency was reinforced by the effects of counterurbanization, such that the polarization of party support between rural and urban areas increased into the 1970s reflecting patterns of class recomposition (Curtice and Steed, 1982).

In the 1979 general election, the Conservatives won all but six of the 72 'rural' constituencies in England and Wales, receiving the support of just under 40 per cent of the electorate (Johnston *et al.*, 1988; Woods, 2002). At a parliamentary level this pattern held throughout the 1980s, although the rural-urban cleavage was blurred by Conservative gains in urban and suburban constituencies. However, a different pattern began to emerge in local government as the Conservatives lost control of a number of rural councils, especially in the south of England, during the 1980s and 1990s, with the Alliance/Liberal Democrats being the main beneficiaries.

The weakening of the Conservative hold over the rural vote in part reflected the weakening of the class cleavage as a determinant of party support. This was manifest not only in a blue collar Conservative vote, but also in the fragmentation of the middle class vote. As Heath and Savage (1992) demonstrate, although the Conservatives increased their support among some middle class occupational groups, such as engineers and the 'bourgeoisie', they also experienced sharp falls in support for the party among other groups, notably public sector professionals such as doctors, dentists, social workers and teachers, as well as scientists and the clergy. Given the significance of public sector employment for service class residents of many more peripheral rural areas this trend in itself helps to explain localized drops in Conservative support, particularly in county towns. Middle class disaffection with the Conservative party was further encouraged by the property slump of the late 1980s. Counterurbanization had stimulated sharp increases in property prices in many rural areas that were not sustained, such that by the early 1990s a significant number of in-migrants to rural areas were experiencing 'negative equity' with their homes worth less than they had paid for them. In 1991, 36.5 per cent of households in central southern England and 33.7 per cent of households in Devon and Cornwall were calculated to have negative equity, and as households with negative equity were unsurprisingly less likely to support the governing party, this can be linked with falls in Conservative support of 2 per cent and 2.6 per cent in these regions respectively between 1987 and 1992 (Pattie *et al.*, 1995).

The ability of first the Liberal-SDP Alliance and later the Liberal Democrats to take advantage of the weakening of the rural Conservative vote was in part a consequence of the national appeal of the centre parties to the service class, but has also been attributed to the persistence of Liberal support in peripheral rural regions, notably south-west England, mid Wales and parts of Scotland. In these regions the Liberal party was able to ride out the collapse in its support nationally due to a combination of distrust of two main parties as parties of the geographical core, a local religious cleavage in which Liberalism continued to be identified with the nonconformist churches and support from small farmers. Between 1950 and 1970, fourteen of the twenty-one seats won by the Liberal party were rural constituencies in these regions. In 1964, an election in which it polled only 11.2 per cent of the vote nationally, the Liberal party achieved 43.4 per cent in the four constituencies of North Devon, North Cornwall, Torrington and Tavistock around the Tamar Valley, over 30 per cent in the rest of Devon and Cornwall, and over 20 per cent in Somerset and Dorset (Tregidga, 2000). Such was the rural concentration of the rump Liberal vote that some figures in the party suggested that it should reposition itself as an explicit 'countryside party', even perhaps adopting the name 'Liberal and Country Party' (Tregidga, 2000). Furthermore, the party had demonstrated its ability to pick up dissident farming votes from the Conservatives at times when Conservative governments had pursued unpopular agriculture policies, with Tregidga attributing the Liberal by-election victory at Torrington in 1958 to this factor.

However, an examination of the local electoral geographies of rural districts suggests that there is a far stronger correlation between the Liberal Democrat vote and experiences of significant social and economic restructuring than there is between the contemporary Liberal Democrat vote and the historic Liberal vote. Put together with the qualitative accounts of the politicization of the service class described in the previous chapter, this evidence suggests a strong anti-establishment element in the realignment of the rural vote. In-migrants and other service class residents who mobilized to defend their own interests and investments during the 1980s and 1990s (see chapter three), frequently did so through the Liberal Democrats because the Conservatives were the party of the existing rural establishment, resulting in the loss of the Conservative hold on rural local government. Many of these voters continued to support the Conservatives in national elections, but the experience of voting Liberal Democrat (or in some cases, Labour) in local elections familiarized them to this possibility and laid the ground for the dramatic shift in electoral support that occurred in 1997.

Indeed, it can be argued that the anti-establishment factor worked against the Liberals in some areas, notably rural Wales. The importance of religious and cultural cleavages in Welsh politics meant that it was the Liberal party that was the party of the rural establishment in much of mid and west Wales, drawing support and leadership from landowners, farmers and business owners who in most other parts of the country would most probably have been Conservatives. Thus five of the eight Liberal MPs elected in 1950 represented constituencies in rural Wales. However, as a new generation of voters emerged and sought to challenge the social and economic conservatism of the local establishment, they turned first to Labour and later to the nationalist party, Plaid Cymru. Madgwick *et al.* (1974), for example, identified clear differences in party support between age groups and occupational groups in Cardiganshire, with the Liberal vote strongly associated with older residents and farmers, and Labour (which won the constituency in 1966 and 1970) drawing support from younger, working class, residents.

The rise of Plaid Cymru as a political force in rural Wales from the late 1960s onwards marked a further generational shift and the mobilization of a community that sees itself as both Welsh and rural. In this way, Plaid Cymru has reproduced a discourse of Welsh national identity that is located in a rural heartland in much the same way as the Conservative discourse of the 1920s located Englishness in the countryside (see Gruffudd, 1995). The Scottish National Party has similarly drawn connections between Scottish national identity and rurality and made many of its first breakthroughs in rural constituencies.

New Labour and the 1997 and 2001 Elections

The only party that has consistently under-performed in rural areas is the Labour party. Despite attempts to present itself as the party of 'rural radicalism' and to mobilize the rural working class, Labour's success in gaining rural constituencies during the twentieth century was limited and the assertion of the class cleavage together with the reproduction of the discourse of the apolitical countryside with its inherent anti-Labour bias effectively relegated the party to a minority position in the electoral politics of the countryside. By the 1980s, Labour's perceived character as an urban and industrial party was largely mirrored in the geography of its parliamentary representation.

Yet, to regain power Labour needed to win at least some rural or semi-rural constituencies, and consequently 'New Labour' made a concerted effort ahead of the 1997 general election to court the rural vote. As Tony Blair's biographer, John Rentoul, records:

Blair had tried to pitch his tent wide enough to take in most of the countryside, being filmed before the election in green Wellingtons, not knowing whether to pat a calf, and giving a transparent interview to Country Life. 'I wouldn't live in a big city if I could help it. I would live in the country. I was brought up there, really'. (Rentoul, 2001, p. 422)

The strategy worked, in so much that rural constituencies swung to Labour with much the same magnitude as the rest of the country. Although the definition of 'rural' constituencies is far from straightforward (see Ward, 2002; Woods, 2002), Labour can be calculated to have polled up to three-tenths of the rural vote in 1997, and took the largest share of the vote in rural Scotland and in constituencies comprised of small towns and their hinterlands (Table 4.1).

Table 4.1 Party support in rural constituencies, 1992 and 1997

	1992			1997		
	Con	Lab	LD	Con	Lab	LD
% of votes cast						
Agricultural areas	49.7	18.8	27.7	38.7	26.9	26.9
Small towns with rural hinterlands	46.9	30.5	20.3	35.5	40.6	17.9
Scottish rural areas	33.4	18.8	22.8	22.9	25.5	22.3
Seats won						
Agricultural areas	44	3	5	35	7	10
Small towns with rural hinterlands	19	4	1	6	16	2
Scottish rural areas	5	3	8	0	5	8

Note: For definition of categories see Johnston et al. (1988).

The Labour party itself used a very broad definition of 'rural' to claim 180 MPs in rural and semi-rural constituencies after the 1997 election, a tally that it further claimed gave it more rural MPs than the Conservatives and Liberal Democrats combined. Whilst the grounds on which this statement are based are open to question (see Ward, 2002; Woods, 2002), it is evident that 1997 witnessed a largely unexpected influx of new Labour MPs representing rural areas, and that this situation caused the government a problem because it had no coherent rural policy to deliver to its new constituents. The Labour manifesto in 1997 had devoted just a couple of paragraphs to rural issues, with the only distinctive policies being commitments to a free vote on hunting and legislation on the 'right to roam'. The challenge of developing Labour's rural policy was taken up by the Rural Group of Labour MPs, formed in 1997 as a response to the mobilization of the Countryside Alliance. One of the first initiatives taken by the group was to commission an academic overview of the social and economic condition of rural Britain, published as the *Rural Audit* in June 1999, which informed the production of a *Manifesto for Rural Britain* in April 2000. At the same time the government charged the Performance and Innovation Unit in the Cabinet Office with examining options for rural policy, leading to the *Rural Economies* report in early 2000. This in turn fed into the government's Rural White Paper for England, published in November 2000.

Collectively, these documents mapped out a distinctive New Labour position on rural policy that was advanced through a three stage process. First, the *Rural Audit* and the *Rural Economies* report provided a description of rural Britain that emphasized the diversity of the countryside, the declining significance of agriculture, the scale and consequences of counterurbanization, and presence of both prosperity and poverty in rural areas. Second, these descriptions were employed to identify the major problems that were judged to be faced by rural areas, with the emphasis placed on issues of health, education, housing, employment, service provision and social exclusion. As such, the discursive terrain of rural policy was shifted on to areas in which Labour has conventionally perceived to be strong, whilst isolating the concerns about hunting, farming and land access that motivated the Countryside Alliance as minor issues. Moreover, they also permitted the assertion to be made that rural areas shared many of the same problems as urban areas, a point articulated in speeches by Tony Blair and by the Agriculture Minister, Nick Brown:

> Taken as a whole, what's striking is how similar the priorities are of those in the countryside and those living in towns. (Tony Blair, speech in Exeter, February 2000).

> When I attend meetings in rural communities, I'm not surprised to find that their aspirations are no different to my constituents in East Newcastle: decent jobs, good schools, a health service that's there when you need it, protection from crime, and an efficient transport system. (Nick Brown, speech to the Labour party conference, September 1999).

Thus, third, New Labour attempted to establish its legitimacy to propose solutions to rural problems. This involved both reiterating the extent of its rural representation and denying the need for any specialist 'rural' expertise because the main problems were shared by rural and urban areas. Indeed, Tony Blair mobilized a 'one nation' discourse to suggest that New Labour was particularly well placed to tackle these issues because it combined both rural and urban representation:

> There are those on the political right who seek to divide town and country, to say that because you live in a city you neither know nor care about those who live in the country. Again, not true. Of course, there are specific rural problems. There are also specific problems in major towns and cities. But there is more that unites us than divides us. There are more common challenges, common values, and indeed common solutions, than there are things that divide us. This is one Britain, one nation, and I will challenge the politics of division wherever they exist. (Tony Blair, speech in Exeter, February 2000).

The political reaction to New Labour's rural policies reveals the new cleavage in rural politics – not a polarization between rural and urban voters, but rather a polarization of voters in the countryside. Rural issues have generated some of the most serious opposition to the New Labour government since 1997. As the remainder of this book will document, issues such as hunting, housing development, the 'right to roam', the handling of the foot and mouth epidemic and the economic situation of agriculture more broadly have all generated conflict that has contributed to the rise of vocal 'rural lobby', manifest in three large scale marches in London, numerous regional demonstrations, and a plethora of 'direct action' protests, including a national fuel blockade. The visibility of this opposition during New Labour's first term of office was such that it was widely anticipated that the party would suffer losses in rural constituencies in the 2001 election.

Yet, analysis of the 2001 election results reveals that any rural backlash against Labour was negligible. Overall, the swing away from Labour in rural areas was not substantially different to the national swing – on several definitions of 'rural constituencies' the swing was significant less than the national average (Woods, 2002). Even in constituencies that might have particularly been expected to exhibit an anti-Labour protest vote – strong hunting areas, seats with significant agricultural employment, the constituencies worst affected by foot and mouth, and low population density areas likely to be hit hardest by fuel price increases – the Labour vote in fact held fairly solid (Table 4.2). Only in the handful of constituencies that suffered the most severe foot and mouth outbreaks was any significant swing from Labour identifiable, and even here the pattern was mixed. Labour increased its vote in six of the twenty worst hit constituencies, including the constituency with the second highest number of cases, Dumfries, and whilst large swings from Labour to the Conservatives were recorded in three Cumbrian constituencies and in Durham North West and Skipton and Ripon, none of these seats were marginal and none changed hands as a consequence. Nationally, the loss by Labour of only one constituency, Carmarthen East and Dinefwr, could conceivably be attributed to rural issues, and even there other local factors were significant (see Woods, 2002, for more).

Thus the two sides of the polarity are revealed. For whilst the traditional rural population that is connected to hunting and farming has become more entrenched in its opposition to Labour, the Labour party has succeeded in appealing to the majority of the rural population that now has no direct connection to agriculture, has little interest in hunting, but is concerned about health, education, housing and employment. As the then chair of the Rural Group of Labour MPs, Peter Bradley, suggested in June 2000, 'one of the reasons I think we may surprise a lot of people is because Labour MPs in rural seats have penetrated those parts of their constituencies the Conservatives never knew existed' (Hinsliff, 2000). This electoral arithmetic means that Labour could afford to ignore rural protests, but also means that it will not be free from them in the foreseeable future. The inability of the Conservative party to now effectively represent rural interests has left direct lobbying and protests as part of an explicitly 'rural' platform as the only course of action for an increasingly embittered and beleaguered 'traditional' rural bloc, as the next chapter discusses.

Table 4.2 Change in Labour vote 1997-2001 by constituency type

	Labour vote 2001	Change 97-01	Labour seats	Change 97-01
One or more case of foot and mouth (93)	36.2%	-2.5	46	=
20 or more cases of foot and mouth (19)	30.6%	-3.3	7	=
Top 20 farming seats (20)	21.4%	-2.3	2	-1
Hunt kennels in seat (159)	30.5%	-1.6	48	-3
Three or more hunt kennels (34)	23.3%	-2.4	6	-1
Car-dependent seats (60)	25.5%	-1.8	5	-2
All GB constituencies (641)	42.0%	-2.2	412	-6

Figures in parentheses are number of constituencies in category.

Source: Woods (2002).

Chapter 5

The Countryside Alliance and Rural Protest

The Emergence of the Countryside Alliance

One of the consequences of the Conservative hegemony was that Britain failed to develop an overt, dedicated, rural lobby. General rural issues were perceived to be represented through the Conservative party, whilst pressure groups such as the National Farmers Union or the Council for the Preservation of Rural England were concerned with particular aspects of the agricultural economy or rural environment and were engaged primarily in sectoral policy-making processes. As long as representation through these largely informal, 'backstage' networks continued to be effective, there was no need for public campaigning by an explicit 'rural' group. However, as the strength of agrarian Conservatism declined, and as the agricultural policy community began to disintegrate, pressure mounted for a switch in tactics.

Among the first to recognize the need for such a change were hunting enthusiasts. The interests of hunting had traditionally been represented by the British Field Sports Society (BFSS), an organization whose leadership tended to be drawn from the ranks of the landed elite and which maintained strong links with the Conservative party. The extent of campaigning by the BFSS aimed at public opinion was limited, and the society instead relied on the use of its supporters' networks to ensure that pro-hunting majorities could be marshaled in parliament to defeat anti-hunting legislation. As the frequency of anti-hunting bills increased in the late 1980s and 1990s, and as public opinion seemed to be strengthening in favour of a ban, some figures within the hunting community became openly critical of the BFSS and its approach:

> The poor quality of the pro-hunting defence has had a shock attack on hunting people. We know this because they have been telling us so by telephone and letter. They are not irked but belligerent. They want something very different – not in two or three more comatose years, but forthwith. (*Hunting* magazine editorial, 1995, quoted by George, 1999, p. 81).

In June 1995, a different strategy was embarked upon with the launch of the Country Sports Business Group, soon renamed the Country Business Group, by an American-born, London-based, lawyer, Eric Bettelheim. The leading figures in the Country Business Group represented a different part of the hunting community to the BFSS – wealthy London business people for whom hunting was weekend leisure activity, rather than a landed rural elite for whom hunting was part of a way of life. The Country Business Group was aimed at one level at fundraising, soliciting donations from businesses with a stake in field sports, but it also sought to apply a business-like strategy to the defence of hunting, believing that hunting had to sell itself to the public if it were to survive. As such, a successful advertising executive, Neil Kennedy, was appointed as director. Despite the different strategy, the Country Business Group saw itself as complementary to the BFSS, not as a rival, and as George (1999) observed the composition of its board suggested 'that new organizations were okay – as long as control rested with "the establishment"' (p. 90).

Five months later, in November 1995, a second new organization was launched, the Countryside Movement. This too was focused on the need to influence public opinion and existed as a non-membership organization with directors drawn from a range of countryside and field sports groups. However, the Countryside Movement set a precedent for future campaigning by realizing that the most effective way to win over public opinion was to campaign on behalf of the 'endangered countryside' as a whole, with field sports positioned as just one of several pressing concerns. In its literature it adopted the tone of an educational initiative, seeking to combat ignorance and misunderstanding about the countryside:

> There is profound misunderstanding of rural life amongst the majority of the population, particularly the young. And there are those who take advantage of this to attack country life, especially livestock farming and country sports. (Countryside Movement leaflet).

The Countryside Movement also broke new ground in recognizing the need to build political links beyond the Conservative party. It appointed the former Liberal leader, Sir David Steel, as its chairman, and a former Labour minister, Lord Cledwyn of Penrhos, to its board of directors, and established connections to the Liberal Democrat Forum for the Countryside, founded around the same time, and sympathizers within the Labour party.

The BFSS, too, began to adopt a more combative and media-aware approach, particularly following the appointment of Robin Hanbury-Tenison as chief executive and Janet George as press officer. Under its new leadership, the BFSS proactively prepared for the anticipated Labour victory in the 1997 general election and the prospect that a bill to ban hunting could be presented to a House of Commons with an anti-hunting majority as early as the summer of 1997. A public rally or march was identified as vehicle through which to demonstrate the strength of support for hunting, but the subsequent organization of this event was strongly influenced by the experiences of two BFSS employees and two key hunt enthusiasts, Charles Mann and Sam Butler, at an anti-gun-law rally held in central London by the Sportsman's Association in February 1997:

> Among the crowd that day were Simon Clarke [BFSS campaign coordinator] and his newly-recruited secretary, twenty-four-year-old Mary Eames. As Clarke put it

afterwards, 'There was a buzz about the event, and at the BFSS we realised that was what we would have to create if we held a rally of our own'. Next day, however, he was dismayed to find that the demonstration received minimal press coverage: tiny, single-column reports in the *Times* and *Daily Telegraph*, and a little bit more in the *Daily Mail*. (Hart-Davis, 1997, p. 4).

Mann and Butler had also joined the Sportsman's rally in February, and thought it brilliantly organised. When they looked back along Picadilly and realised that half of the thoroughfare was a solid mass of marchers, as far as they could see, they thought, 'Fantastic! This must be bringing London to a halt'. Yet when they slipped out of Trafalgar Square with speeches still in progress, and went round one corner, they found everything perfectly normal. 'London didn't know what was happening', Mann recalled afterwards, 'We saw that over 20,000 people could march through the streets, and London didn't notice'. (Hart-Davis, 1997, p. 5).

The rationale was hence developed that in order to derail any parliamentary effort to ban hunting, the hunting lobby had to gain significant media attention; that in order to gain media attention it needed to attract substantially more than 20,000 people to a demonstration in London; and in order to attract large numbers to a demonstration, it had to campaign on broader 'rural' issues, not just on hunting. As Hart-Davis records, Charles Mann had advised Simon Clarke that 'in any initiative the "countryside aspect" must be emphasised, because if a rally were seen to be for hunting only, many people would be alienated' (Hart-Davis, 1997, p. 5).

The proposed rally was rebranded as the 'Countryside Rally' and promoted as a demonstration by rural people about a range of issues. The message of the 'countryside' mobilizing as a united force was further conveyed by the merger of the BFSS, Country Business Group and Countryside Movement to form the Countryside Alliance in March 1997. Although the merger was motivated partly by financial concerns and partly by the need to overcome the BFSS's 'Tory party on horseback' image (George, 1999), it also had the benefit of removing the explicit reference to field sports and emphasizing the 'rural' identity of the group. Indeed, although at the time of the 1997 Countryside Rally the three groups still maintained separate legal identities and had not yet fully merged, and although the burden of organization of the event fell to the BFSS, the name of the BFSS was deliberately left off promotional literature even though it was the society's contact details that were given (Hart-Davis, 1997).

The Countryside Rally and Marches

The first Countryside Rally was held on 27 July 1997, a date selected as the first opportunity on which a private members bill to ban hunting could have been presented to a parliament elected on 1 May. Over 120,000 people attended the rally in Hyde Park, brought from across the country by 924 chartered coaches and 12 chartered trains. The rally was addressed by a collection of politicians, writers, celebrities and representatives of field sports organizations, and was followed by the presentation of a petition to Downing Street. The rally was also the end-point for 149 long-distance marchers who had followed five routes from Machynlleth and St Clears in Wales, Caldbeck in Cumbria, Coldstream in Scotland and Penzance in Cornwall, wearing t-shirts and sweatshirts with the slogan 'Listen to Us' (Hart-Davis, 1997). In broad terms, therefore, the strategy behind the Countryside Rally proved remarkably successful. Not only had the event attracted one of the largest crowds of recent years, but it gained extensive media attention. Over 300 television and radio interviews were given from the rally and several pages of coverage were achieved in most national newspapers (George, 1999). Furthermore, although hunting featured prominently in the speeches given to the rally, the media largely represented the demonstration as a 'countryside' protest, not just as a pro-hunting protest. The political impact was two-fold: it forced concessions from the government, particularly with-holding government time in parliament from an anti-hunting bill, and it placed rural issues at the heart of the political agenda.

The momentum was maintained by the Countryside March on 1 March 1998. The date was once again significant, this time corresponding with the report stage of Michael Foster's private member's bill on hunting, but encouraged a change in format. As Janet George recalls, 'Park authorities would not permit a Rally in Hyde Park during the winter months, and standing around for hours on wet or icy ground would be too unpleasant to contemplate anyway' (George, 1999, p. 140). Instead the 250,000 participants marched through central London from Embankment Station to Hyde Park, where they were counted and dispersed without speeches. The message of the march was hence conveyed through the placards carried by participants and deposited at the entrance to the park. Although most related to hunting, a minority alluded to other issues including the agricultural slump (see chapter six), housing development and the closure of rural services. As such they exhibited the tension that ran throughout the organization of the marches between the desire to engage with a range of rural issues and the core motivation of hunting. Janet George, in her account of the march, recalls the deliberate meshing together of these concerns:

> Michael Foster's bill was the focus for the March, but the 'countryside' theme was promoted al the way. Access and the risks of a statutory 'right to roam' were coming up for debate, and farmers' problems were escalating. With hunting completely dependent on the goodwill of farmers and landowners, it was an excellent opportunity for a huge display of unity. (George, 1999, p. 145).

A further march was planned for 2001, but cancelled due to the Foot and Mouth crisis. In September 2002, the Countryside Alliance organized its third major London demonstration with the Liberty and Livelihood March. With over 400,000 participants, the march not only exceeded the size of the previous events, but also demonstrated the increased professionalism and sophistication of the Alliance as well as the increasing militancy of its supporters, as discussed later in this chapter. Whereas the earlier demonstrations had kept away from parliament, the two routes of the Liberty and Livelihood March, starting at Hyde Park Corner and Tower Hill respectively, converged on Whitehall and Parliament Square, with a strategically positioned media platform ensuring that the marchers were filmed dispersing in the shadow of Big Ben. Both the 1998 and 2002 marches were preceded by the lighting of beacons around the country, providing the media with striking images and contributing to publicity in the build-up to the marches that ensured several days of news coverage. Similarly, the distribution of plastic signs promoting the marches around the country reinforced the impression of the 'countryside coming to town'.

The ability of the Countryside Alliance to organize three major public protests in London demonstrated both its rapid emergence as one of Britain's leading pressure groups and the substantial financial and professional resources at its disposal. The 1997 rally cost £436,810 to organize, the 1998 march, £450,071 and the 2002 march, £327,444 (George, 1999; Countryside Alliance, 2003). Whilst the Countryside Rally made a loss, the cost of the two later marches was more than covered by donations, although the Countryside Alliance as a whole operates in deficit. In 2002, the Countryside Alliance had an income of just under £7.5 million, of which £2.82 million came from subscriptions, £1.46 million from fundraising, and just over £3 million from donations. Its expenditure in the same year was £8.3 million, of which around a quarter was spent directly on campaigning (Countryside Alliance, 2003). Professional resources are, however, as important to the Countryside Alliance as finance. The organizing teams for the London demonstrations brought together professional expertise from the media, public relations, event management and the military. Moreover, the Alliance's capacity to mobilize large numbers of people from across the country has relied not only on its own regional teams but also on the voluntary support network provided by local hunts. Over half of the 2,244 coaches that took participants to the Liberty and Livelihood March were chartered by hunts or hunt supporters clubs (Tables 5.1 and 5.2).

Table 5.1 Chartered coaches to the Liberty and Livelihood March by organizer

Hunts and hunt supporters clubs	1282
Gun clubs, wildfowling clubs and shoots	269
Estates	173
Individuals	140
Countryside Alliance branches	40
NFU / FUW / Young Farmers branches	22
BASC branches	5
Conservative Associations	2
Other	286
Not known	19
Total	2244

Source: Liberty and Livelihood March Commemorative Magazine.

Table 5.2 Liberty and Livelihood March, chartered coaches, passengers on chartered trains and beacons by county

	Coaches	Train*	Beacons**
Bedfordshire	8	1120	20
Berkshire	47	-	20
Buckinghamshire	29	-	20
Cambridgeshire	34	-	30
Cheshire	41	710	20
Cornwall	36	-	50
Cumbria	39	-	20
Derbyshire	35	-	60
Devon	129	820	50
Dorset	96	-	60
Durham	15	600	30
East Sussex	18	1650	20
Essex	55	-	20
Gloucestershire	171	-	289
Hampshire	125	1870	40
Herefordshire	50	-	35
Hertfordshire	25	900	20
Isle of Wight	-	-	50
Kent	12	3780	20
Lancashire	45	-	20
Leicestershire	99	-	110
Lincolnshire	44	-	20
Norfolk	110	50	20
Northamptonshire & Rutland	45	-	60
Northumberland	12	1090	20
Nottinghamshire	40	600	60
Oxfordshire	106	1096	160
Shropshire	56	-	50
Somerset	84	700	50
Staffordshire	56	-	30
Suffolk	110	-	20
Surrey	125	-	30
Warwickshire	69	-	60
West Midlands	2	-	-
West Sussex	60	250	30
Wiltshire	107	1336	200
Worcestershire	73	480	50
Yorkshire	137	825	50
Scotland	10	-	30
Wales	163	-	60

* Number of passengers booked on chartered trains originating in county
** Approximate number of beacons in county
Source: Liberty and Livelihood March Commemorative Magazine.

The three London demonstrations have been the Countryside Alliance's most high profile initiatives but represent only a fraction of its activities. In the periods between the major London marches, demonstrations have been organized at a local and regional scale, mobilizing displays of strength throughout Britain. In the autumn of 1999, for example, rallies in Birmingham, Newcastle, Norwich, Exeter and Cardiff all attracted between 10,000 and 20,000 participants. Demonstrations have been regularly organized outside the Labour party conference, and also at meetings of other anti-hunting groups such as the RSPCA. The separate threat by the Scottish Parliament to ban hunting, was met with (unsuccessful) protests in Edinburgh, including the 'March on the Mound' in December 2002. Equally important has been the development of more conventional political lobbying focused on Westminster and the media. For example, in 2002 the political department circulated 55 briefings to MPs and peers, the policy development team responded to more than 25 government consultations, whilst the public relations department handled over 10,000 enquiries from the media and issued 120 press releases (Countryside Alliance, 2003). The activities of these departments also illustrate the broadening of the Alliance's platform. Although the 'Campaign for Hunting' remained its flagship activity, the Alliance also launched a 'Campaign for Shooting', a 'Campaign for Angling' and a 'Racing Initiative'. Significantly, the Countryside Alliance also initiated a more practical engagement with rural development issues. An 'Honest Food' campaign involved promoting farmers markets and sources of locally produced food and a competition for pubs that sourced food and drink locally; a seminar on rural employment was organized in 2002 and a conference of key rural organizations in the same year; and a Rural Regeneration Unit was established in 2003 as a self-financing not-for-profit company that would 'assist relevant government departments and agencies, public sector bodies and charitable foundations in translating public policy into grass-root action and self-help' (Countryside Alliance, 2003, p. 9). In an annually-updated Policy Handbook, dubbed 'The Real Rural Agenda', the Countryside Alliance has set out its policies not just on country sports but also on agriculture, animal welfare, rural enterprise, housing, employment and poverty, conservation, tourism, rural services, transport, crime and policing, health care and education.

At a local level, the Countryside Alliance operates through fifteen regions, with regional committees and regional directors. As well as working with the regional media and organizing local campaigning events, the regions also support a range of social activities. These are both a significant source of fundraising for the Alliance and work to position the Alliance as part of the social infrastructure for its core constituency. To facilitate its work across all these activities, the Countryside Alliance employed 85 salaried staff at the end of 2002. Whilst the expansion of the organization's work has generated tension among its supporters, as is discussed later in this chapter, it has been fundamental in positioning the Countryside Alliance not just as a protest group but as a legitimate vehicle for the representation of rural interests, a status that has been recognized in its incorporation in a number of quasi-governmental fora and partnerships.

The Rural Movement

The rise of the Countryside Alliance is part of a broader emergence of a rural social movement that has taken place not only in Britain, but also in many other developed nations (Woods, 2003a). The conditions for the mobilization of the rural movement were set by the destabilizing of existing social and economic structures by rural restructuring and by political developments that weakened the capacity of conventional rural pressure groups and elites to deliver results. Thus, in seeking to defend their traditional communities and ways of life, an increasingly beleaguered 'rural' population has turned to new political organizations and new tactics to get their message across. This has produced not only the Countryside Alliance, but also a number of smaller protest groups in Britain (see later in this chapter), as well as a plethora of other organizations around the world, including political parties such as the *Chasse, Pêche, Nature et Tradition* (Hunting, Fishing, Nature and Tradition) party in France, conventional lobby groups such as the Rural Coalition in the United States and Mexico, and campaign movements engaged in protest activity, including the *Confédération paysanne* in France, the Family Farm Defenders in the United States and the *Platform Buitengebeid* in Belgium. That is not to say, however, that there is a coherent, coordinated platform presented by these groups. The organizations that might be identified under the loose description of a 'rural movement' take many different organizational forms, practice different strategies and tactics, are founded on different ideological positions and advocate a range of, sometimes contradictory, policies. In this way the constellation of rural campaign groups resemble a new social movement, characterized by della Porta and Diani (1999), following Offe (1985), as involving,

> an open, fluid, organization, an inclusive and non-ideological participation, and greater attention to social than to economic transformation. (della Porta and Diani, 1999, p. 12).

As della Porta and Diani further argue, drawing on Gerlach (1976), new social movements are *segmented*, 'with numerous different groups or cells in continual rise and fall' (p. 140), *policephalous*, 'having many leaders each commanding a limited following only' (ibid.), and *reticular*, 'with multiple links between autonomous cells forming an indistinctly bounded network' (ibid.). The lack of coherence of the rural movement is therefore in fact part of its identity as a new social movement, distinguishing it from older forms of political organization. So too, is the fact that the primary motivation for the groups involved is not the furthering of economic or material interest, or class action, but the defence and promotion of a 'rural identity'. Whilst different groups within the rural movement might define rurality differently, it is nonetheless their common reference point (see also Woods, 2003a).

The location of the Countryside Alliance as part of the rural movement is based on three features of its organization and activity in which it reflects the attributes of the rural movement as a new social movement. The repertoire of tactics and strategies employed by the Countryside Alliance parallel those of other social movement organizations in ranging from formal lobbying and the use of the legal and policy processes through to publicity stunts and demonstrations that verge on civil disobedience. Furthermore, whilst the Countryside Alliance is at the more restrained, conservative end of rural movement, there is a degree of fluid movement of participants in its events and in more radical, sometimes illegal, direct action under the auspices of other rural protest groups. In common with most social movement organizations, a large proportion of the Countryside Alliance's activities are primarily aimed at gaining media coverage. As della Porta and Diani describe,

> For the most part social movements use forms of action which can be described as disruptive, seeking to influence elites through a demonstration of both force of numbers and activists' determination to succeed. At the same time, however, protest is concerned with building support. It must be innovative or newsworthy enough to echo in the mass media, and, consequently, in the wider public which social movements (as 'active minorities') are seeking to convince of the justice of their cause. (della Porta and Diani, 1999, p. 183).

The rationale for the broadening of the Countryside Rally strongly resonates with this statement, as does the logic behind many of the Alliance's subsequent activities. As discussed further below, the three mass demonstrations in London were all highly orchestrated events aimed at delivering clear, symbolic messages. Symbolic publicity stunts aimed at attracting media attention have also been used at other smaller events, including the presence of topless female riders bodypainted in hunting outfits at a demonstration outside DEFRA offices. The orchestration also involved the discursive manipulation of space. The Countryside Rally and London marches were richly imbued with the symbolism of 'the country comes to town' and followed carefully selected routes through key landscapes of nation identity, such as Trafalgar Square and Whitehall, and sites associated with political dissent, including Hyde Park and, again, Trafalgar Square. In these ways, the Countryside Alliance can be argued to have engaged with a 'postmodern politics of resistance', that, as Routledge describes,

> mounts symbolic challenges that are extensively media-ted in order to render power visible and negotiable, and to attract public attention. Such a politics, and the spaces within which, and from which, it is articulated are frequently hybrid in character and ambiguous in practice and effect. (Routledge, 1997, p. 372).

Secondly, whereas traditional political organizations, including the established rural interest groups, were hierarchical in nature, with the leadership delivering to a largely passive membership, new social movement organizations, including the Countryside Alliance, depend on the active participation of their members. The Countryside Alliance claims to have over twenty different means of communicating with its membership, including e-mail newsletters that enable the stimulation of rapid responses. The main e-newsletter, the 'grass e-route' is estimated to have a reach of around 250,000 people, partly through its inclusion on 500 other websites, and is supported by 14 regional e-newsletters with a combined mailing list of around 35,000 names (Lusoli and Ward, 2003). All the methods of communication actively encourage the Alliance's members and supporters to participate in national and regional protest events, write letters and sign petitions, and contribute to media debates, phone-ins and online polls. A programme of social and fundraising activities further reinforces the engagement of the membership and the blending of social identity and political purpose means that the division between 'social members' and 'political members' that exists in many political parties and some established pressure groups is not evident in any significant way in the Countryside Alliance. Instead there is a high degree of involvement of members in both social and campaigning activity (Table 5.3).

Table 5.3 Involvement of Countryside Alliance members in campaigning and social activities

Read Countryside Alliance literature	73%
Attend rallies and/or demonstrations	60%
Talk to colleagues/friends about Countryside Alliance	51%
Donate money	49%
Attend fairs / social events	36%
Meet with other members	14%
Campaign for the Countryside Alliance	10%
Hold an official position in the Countryside Alliance	2.5%
Visit Countryside Alliance offices	2%
Volunteer for clerical work	1.5%

N = 1190

Source: Lusoli and Ward (2003).

Thirdly, the Countryside Alliance reflects one of the key characteristics of new social movements in defining its constituency in terms of rural identity. Whereas traditional interest groups and 'old social movements' tended to be based on the defence or promotion of material interest, Fainstein and Hirst (1995) have argued that 'the "new" social movements cut across classes and are guided by non-material considerations' (p. 183). In the rural context this has meant a transition from organizations that represented farmers or landowners or agricultural workers – discrete groups defined by their economic interests – to organizations that claim to represent a 'rural identity' (Woods, 2003a). The Countryside Alliance, in seeking to mobilize a broad movement of 'rural people' from different classes and different economic sectors, is one of the clearest examples of this practice.

Yet, the socially constructed and contested nature of rurality means that whilst different groups may claim to be mobilized around 'rural identity' their understanding of that term can vary considerably. Within the rural movement as a whole, at least three distinct strands of rural identity can be identified – a *reactive ruralism*, involving the mobilization of a self-defined 'traditional' rural population in defence of purportedly historic, natural and agrarian-centred 'rural ways of life' in response to a perceived challenge from 'ill-informed' urban intervention; a *progressive ruralism*, involving action in opposition to modern farming practices and agricultural policy, the globalization of trade in food, land-ownership patterns, road developments or other activities that conflict with a discourse of a simple, close-to-nature, localized and self-sufficient rural society; and an *aspirational ruralism*, involving the mobilization of in-migrant and like-minded actors to defend their fiscal and emotional investment in rural localities by seeking to promote the realization of an imagined 'rural idyll', and resisting developments that threaten or distract from this idyllic countryside (Woods, 2003a). On certain issues actors from these different strands are able to work together because there is a correspondence of their rural visions – on road-building and housing development, for example (see chapter seven), or on the promotion of local foods. On other issues, however, the different ruralisms lead to diametrically opposed positions such that their protagonists are drawn into conflict with each other.

With its roots in the hunting lobby and mobilization in response to the election of the Labour government in 1997 and the anticipated threat of anti-hunting legislation, the Countryside Alliance clearly represents an example of reactive ruralism. There are particular issues and campaigns on which the Countryside Alliance is able to make common cause with other organizations that primarily represent other strands of ruralism – in its 'Honest Food' campaign, for example – and the breadth of the Alliance's campaigning and rhetoric may well have enrolled individual members who would personally identify with a progressive or aspirational ruralist position, yet, at its core the organization is founded on a reactive ruralism. This is evident in the definition that the Countryside Alliance itself offers for the 'countryside' that it represents. In this, the countryside is defined not in spatial or material terms, but through lifestyle and attitude:

> The Countryside Alliance believes the countryside is best defined by its inhabitants. Families involved in traditional, conservation-minded farming and allied trades are part of the true rural population. So too are people who participate in country sports, and support and identifiable rural culture and rural system of values. This includes many recent settlers from towns, as well as many who, by circumstance, are forced to live in towns and cities for at least part of their lives. (Countryside Alliance website, www.countryside-alliance.org.uk).

The exclusionary tenor of this definition – referring to a 'true rural population' – was replicated in contributions made to an interactive discussion on the Countryside Alliance's website in early 2002 which addressed five questions including 'what is rural?' and 'what do you consider the characteristics of being rural?'. Whilst the messages posted to the bulletin board revealed a range of perceptions among contributors, and included relatively inclusive and geographical definitions, many alluded to a requirement for rural people to adhere to particular practices and attitudes, and even explicitly excluded who were considered not to fit:

> 'Rural' is as much a state of mind as an actual place. It is an acceptance and understanding of people and things living in a mainly agricultural area, the practices and traditions. (Posting to Countryside Alliance website, www.countryside-alliance.org.uk/policy/whatis/index.html).
>
> Living and working in the countryside – with roots in the countryside from childhood. An understanding of the countryside and an unsentimental attitude to the animals, both wild and domesticated. (*Ibid*).
>
> Having a clear understanding of country life, of the strong links between farming and outdoor pursuits and understanding the true values of living in the countryside. This is opposed to someone moving there and growing geraniums in hanging baskets and putting fake farm implements in their garden to make it look authentic, at the same time, out pricing genuine 'natives' out of the housing market. (*Ibid*).
>
> Having an in-depth knowledge and understanding of the countryside and country matters. Simply moving into and living in the country does not qualify. (*Ibid*).

Perhaps most powerfully, the essence of a reactive ruralism was articulated through the three large demonstrations in London – the Countryside Rally of 1997, Countryside March of 1998, and the Liberty and Livelihood March of 2002 – and in the accompanying publicity and media commentary. As the next section describes, these events reproduced a defensive vision of the countryside under threat from urban interference that resonated strongly with the traditional conservative discourse of rurality.

Reading the Countryside Protests

As described above, the major protests organized by the Countryside Alliance have been orchestrated events, carefully designed and planned to communicate a particular message to a particular audience. The Countryside Rally in July 1997, the Countryside March in March 1998 and the Liberty and Livelihood March in September 2002 may all, therefore, be 'read' as performances – staged and scripted with a physical materiality and visual presence. The communication of the message, however, relied not only on the visual consumption of the protests themselves, but also on a series of peripheral stunts, such as the lighting of beacons around the countryside, publicity material and commemorative souvenirs, and media briefings, that sought to direct the interpretation of the protests. Thus the surrounding commentary both connected the protests to larger political themes and ideas, such as the defence of liberty and the assertion of a traditional sense of British national identity, and represented the rallies and marches as performances of a particular rural identity which again was founded in a very traditional discourse of rurality. Taking all these elements together, it is possible to examine the ways in which the countryside were constructed by their organizers and media supporters, and then to step back and to critically deconstruct the protests to problematize the representation of the countryside that they conveyed.

Constructing the Countryside Protests

The strategic decision of the pro-hunting lobby to recast themselves as broad movement defending the interests of the countryside was based on a rational assessment of the political situation and their chances of exercising influence. However, the evolution of the arguments and representations subsequently employed by the Countryside Alliance was itself part of the process through which the lobby sought to rationalize the problem confronting it – constructing its own 'reality' about the nature of the contemporary countryside and its political context, which flowed from the interests of power vested in the rural movement. As Flyvbjerg (1998) observes, 'power *defines* what counts as rationality and knowledge and thereby what counts as reality' (p. 227), thus the argument mobilized by the Countryside Alliance was prevaricated on a series of assumptions and rationalizations.

Firstly, the hunting debate was constructed as a rural-urban conflict, or rather, as part of wider rural-urban political struggle. This rationalization not only involved an implicit definition of who and what is 'rural', but also established the fault-line that structured the subsequent development of the campaign. By identifying the rural-urban divide as the dominant fault-line, a commonality was suggested between hunting and other problems faced by rural areas. Thus, problems of rural poverty and unemployment could be represented as having more in common with the threat to hunting, than with problems of poverty and unemployment in urban areas, and so on – legitimizing the creation of a broad-based 'countryside' protest movement. The rationalization of a rural-urban conflict positioned misgovernment by the urban population as the cause of social and economic ills in the countryside and hence demanded solidarity amongst rural people in protest against the urban hegemony, to assert their 'independence'.

In the 'reality' of the Countryside Alliance the rural-urban division was not only geographical but also cultural and moral. Thus the political problem was not just a question of geographical representation or self-government (although the Countryside Alliance ingeniously pointed out that rural areas accounted for 75 per cent of Britain's land mass but just 11 per cent of Members of Parliament (Hanbury-Tenison, 1997)), but stemmed from the incompatibility of rural and urban cultures. This constructed dichotomy was eagerly reproduced by the media as an explanation of the countryside protests, for example by Henry Porter in *The Guardian*:

> In the country, people talk of property and farms; in the town, of nature and the outdoors. The first contains ideas of work and ownership, the second of appreciation and enjoyment. (Porter, 1997, p. 2).

Secondly, however, in the discourse of the countryside lobby, the dichotomy of rural and urban cultures was not equally balanced. The rural was rationalized as forming a beleaguered minority under attack from an insensitive urban majority, as articulated in the Countryside Rally's mission statement:

> This initiative arose as a response to the frustration and concern felt by country people against the threats posed to the countryside and their jobs, by politicians and urban influence, through prejudice, ignorance and diminishing rural representation. (quoted by Porter, 1997, p. 2).

Despite gaining a significant number of rural constituencies at the 1997 general election (see chapter four), the Labour party was represented in this 'reality' as an 'urban government', whose values and policies were rooted in urban culture and thus which had no understanding of rural issues – a charge conveyed by the Conservative party leader, William Hague, in the *Daily Telegraph* before the Countryside March:

> I suspect that the reason why the Government is behaving so badly to rural Britain goes deeper. It is based on a complete ignorance of life in the countryside as it is actually lived. (Hague, 1998).

Moreover, whilst urban culture was represented as being numerically dominant, rural culture was represented as being morally superior. Thus whilst the director of the Countryside Alliance wrote of 'the sweet voice of rural reason' (Hanbury-Tenison, 1997, p. 92), the actions of 'urban politicians' in seeking to ban hunting were portrayed as being based on 'ignorance' and 'prejudice'. The *Daily Telegraph*, in a leader column defence of hunting, lamented that:

> something that has worked and been much loved for centuries is told that it must now be abolished because of an urban, anti-Christian ideology which sees animals as being in no moral sense different from human beings and prudishly refuses to confront the reality of life and death in nature. (*Daily Telegraph* leader, 28 February 1998).

Whilst the pro-hunting Labour peer and President of the Countryside Alliance, Baroness Mallalieu, told the Countryside Rally:

> We cannot and will not stand by in silence and watch our countryside, our communities and way of life destroyed forever by misguided urban political correctness. (quoted in the *Daily Telegraph*, 11 July 1997, p. 4).

These representations draw heavily on the long-standing anti-urban discourses in which urban culture is dismissed as corrupt, decadent and unpure (see chapter four). The resonance of these discourses in the countryside lobby's arguments was explored by David Aaronovitch, who, caricaturizing the lobby's rationalization of the rural-urban divide, wrote:

> the city is degenerate, addicted to fashion, a sink of vice, a destroyer of health and corrupter of morals; it makes men effete and women adulterous. Removed from any

connection with a 'natural' world that it cannot understand, it nevertheless reaches out tentacles of pollution and development to destroy the peace and happiness of Arcadia. (Aaronovitch, 1998, p. 21).

In contrast, Aaronovitch continued, within the discourse of the countryside lobby, the countryside was a land in communion with nature, whose 'fields and villages preserve the traditions and the heritage of the nation' (*ibid*.). Whilst the 'arrogance' of the urban in attempting to impose its cultural values on the rural was represented as an act of 'majority dictatorship', with the *Horse and Hound* magazine distributing placards for the Countryside March with the slogan 'Say No To The Urban Jackboot', the countryside was constructed as the guardian of true freedom.

Thus, the countryside protests were rationalized by their organizers and participants as being primarily about fighting for freedom. This operated on three levels. At one level the attempt to legislate to ban hunting was represented as an unwarranted intervention by the state into the private sphere which infringed civil liberties. At a second level, the right of an 'urban majority' to legislate on issues which concerned mainly rural people was challenged as illiberal and undemocratic. This was not presented as a call for devolution, but rather as an assertion that British democracy depended on the majority *listening* to the minority:

This rally is not just about hunting ... It is about freedom, the freedom of people to choose how to live their own lives. It is about tolerance of minorities – and, sadly, those who live and work in our countryside are minorities here today. And above all it is about listening to and respecting the views of other people of which you may personally disapprove. (Baroness Mallalieu, quoted in Hart-Davis, 1997, p. 127).

Moreover, at a third level, the countryside protests were represented as performing the countryside's historic role as the repository and defender of national freedom and values. The function of the Countryside Rally, Countryside March and Liberty and Livelihood March, it was implied, was to correct the transgressions of the urban elite:

The countryside has come to London to speak out for freedom. (Baroness Mallalieu addressing the Countryside Rally, quoted in Hart-Davis, 1997, p. 128).

It makes me angry that so many people will have to travel so far to ask our Government – a Labour Government – to defend our freedom to choose for ourselves how we live our lives. (Baroness Mallalieu, in the *Liberty and Livelihood March Commemorative Magazine*, 2002, p. 15).

Intrinsic to these representations was the entwining of rurality and national identity. In particular, the countryside lobby drew both implicitly and explicitly on historical positioning of the rural at the core of English national identity (see chapter four), to suggest that the perceived threats to a rural way of life were also threats to the fundamental English way of life. This association was articulated by the novelist Joanna Trollope, writing in the *Sunday Telegraph* on the day of the Countryside March:

> There is something singular about the relationship between the English and their countryside ... a robust instinct for living in practical harmony with nature (resulting in that much admired social unit, the English village), and an almost pantheistic appreciation of landscape. (Trollope, 1998, p. 1).

> the English continue to feel a determined union with the countryside. It is a sense of both belonging and finding salvation there, in a community – preferably consisting of a church, pub, farms, cottages, a small school and a Big House. We have, we English, a national village cult: we cherish the myth that out there, among fields and woods, there still survives a timeless natural innocence and lack of corruption. (*Ibid.*).

Trollope continued to describe the processes of social and economic change that have begun to threaten this rural idyll. Yet she ends defiantly:

> powerful as these forces are ... they will not alter the fundamental feeling that the English have: that it is their innate, inalienable right to live in the countryside ... the crowds in Hyde Park today are not marching just for themselves and their way of life; they are marching for the nation. (*Ibid.*).

Or as the *Daily Telegraph* leader column concurred:

> The people who are coming to London are the backbone of the nation. They are those who have always been ready to fight for their country when required. For them 'country', in the sense of nation, is closely bound up with 'country', in the sense of green fields. (*Daily Telegraph* leader, 28 March 1998).

Here we have the final and most powerful rationalization of the countryside lobby's position. The countryside protests are not just about defending hunting, nor just about representing rural interests, nor just about standing up for freedom and civil rights, they are about defending the nation, standing up for true national values. They are not directed at the urban population, but at an amorphous un-English, un-British, urbanity – they are about saving urban people from themselves.

Thus, both Charles Moore, the editor of the *Daily Telegraph*, and William Hague, the Conservative party leader, explicitly identified the rural values of the countryside protesters as being *true British values* (as opposed, it was implied, to the un-British values of the urban elite):

> I was not marching with sadists yesterday, but with tens of thousands of good, true British people ... We are dealing with an aspect of the British character which is common to all classes. This is a phenomenon which has led our country to win wars. It is summed up in the phrase 'Leave us alone'. (Moore, 1998, p. 20).

The reason why Sunday's march is so important is that the Government's attack on the countryside is also an attack on many peculiarly British values, which city-dwellers feel as keenly as rural dwellers. (Hague, 1998).

As such, Hague argued, echoing the discourse already mobilized by Trollope and others, the protesters were protesting not for themselves, but for all true patriots:

If you believe that the particular British ability to change gradually and peacefully contributes to the quality of our life; if you believe that institutions should grow organically and not be imposed according to the latest blueprint, however 'cool'; most of all, if you believe that your life is your responsibility, and not that of a minister or civil servant, then you should know that the Countryside Marchers are marching for you as well. (Hague, 1998)

The patriotic mission of the countryside protests was underscored by the appropriation of national symbols and signifiers. Many participants in the Countryside March and the Liberty and Livelihood March carried union jacks, although in a blurring of the national identities being evoked, Welsh, Scottish and Cornish flags and the standard of St. George were also numerous (Figure 5.1). A union jack adorned the front the bus conveying a delegation from the Countryside Rally to Downing Street to present a letter of protest; whilst the national flags of Wales, Scotland and Cornwall were carried at the head of the 1997 marches. Baroness Mallalieu recited of the eve of battle speech from *Henry V* to the Countryside Rally, evoking both the romanticized Englishness of the Shakespeare cult, and the militaristic nationalism conveyed in the prose. Similarly, the beacons lit across rural Britain before the Countryside March and Liberty and Livelihood March deliberately resonated with the beacons historically lit to warn of invasion, and, in a more contemporary context, the beacons that marked the nationalistic celebrations of royal jubilees and weddings and the anniversary of VE Day. The Hyde Park crowd at the Rally sang the patriotic hymn, *Jerusalem* - beloved of the Conservative rural-nationalists of the inter-war years - and *Jerusalem* was played, along with other patriotic music including *Land of Hope and Glory*, Souza's *Dambusters* theme, *Rule Britannia*, *Flower of Scotland*, *Danny Boy* and *Land of our Fathers*, on the March FM radio station during the Countryside March – leading a *Guardian* journalist to comment that it suggested 'that you can only be truly British if you live in the countryside and like to kill animals' (Karpf, 1998, p. 5).

Equally, the countryside protests were positioned by their protagonists in a traditional of radical protest. The Jarrow March of 1936 formed an important inspiration for the marches to the Countryside Rally and later the Countryside March; whilst the whole series of protests were compared by their organizers to the Peasants Revolt of 1381 (Hart Davis, 1997). The *Daily Telegraph* went further, placing the Countryside Rally in a list of British protest movements including Peterloo, the Jarrow March, the Aldermaston March and the Bloody Sunday demonstrations in Londonderry. A similar litany was presented by the director of

Figure 5.1 Flags on display in the Liberty and Livelihood March

the Countryside Alliance's Campaign for Hunting in the commemorative magazine for the Liberty and Livelihood March, which ended with an even more ambitious comparison:

> We follow in the footsteps of some distinguished campaigners who, with a mixture of patience and resolve, achieved their objectives. Mahatma Ghandi and 78 friends had to march 241 miles from Sabarmati Ashram to Dandi, Gujarat, to highlight the inequities of the Government levy of salt tax, a gesture that has formed the benchmark for all subsequent libertarian campaigning. (Hart, 2002, p. 39).

This heritage was further evoked by the spatial significance of the site of the Countryside Rally and the routes of the two marches. Not only was the act of 'the countryside coming to town' symbolically important in claiming space at the heart of the nation; but the Rally was deliberately located at 'The Reformer's Tree' in Hyde Park, a symbol of British democratic tradition, and the marches' crossing of Trafalgar Square invited identifications with both the nationalistic iconography of the site, and with previous demonstrations.

Yet, the tradition of radical protest in Britain assumed by the countryside lobby was one shaped by its nationalistic, Conservative ideology (see chapter four). Whilst favourable evocations were made of protests which could be represented as organic uprisings of the British people against the state – the Peasants Revolt, the Jarrow March etc. – and which therefore could be positioned as elements in the construction of British national values; the tradition clearly did not include more recent protests – against the Poll Tax, for Gay rights – which were regarded as transgressing those same British values. So, the *Daily Telegraph* mentioned the anti-Poll Tax demonstrations, but emphasized the violence and rioting which accompanied them, and even devoted space on its front page to asserting that 'Clapham Common may have required days of rubbish gathering after Saturday's Gay Pride march but just an hour after yesterday's rally, litter was scarce and Hyde Park's grass looked its normal self'. Similar distinctions were drawn by Charles Moore after the Countryside March:

> We have grown so used to rent-a-mob that we have forgotten what a genuine mass demonstration is like ... [The march was] certainly the most respectable public protest in British history. (Moore, 1998, p. 20).

In disassociating the countryside protests from recent political demonstrations, the countryside lobby again drew upon another fundamental aspect of the traditional discourse of British rurality – that the countryside is an apolitical space, where enduring traditions and values are more important that urban ephemera such as politics. Thus any protest in defence of those rural values was also therefore not a political act. This assertion produced confusion in the attitude of government agencies towards the protests. Farmworkers in Berkshire were prevented from lighting a protest bonfire on land owned by the government Institute for Animal Health because it was deemed a 'political act', but the Ministry of Defence defended its donation of VIP tours of RAF Leeming in

Yorkshire as prizes in a fundraising raffle for the Countryside March as improving relations with the local community with 'no political overtones whatsoever'. The discourse was reproduced by the Countryside Alliance and its supporters, who dismissed criticism of the protests as 'politically motivated'. In contrast, it was implied, the protesters were motivated not by politics, but by a deeper, more instinctive, national instinct:

> This is not a subversive protest. This is a protest by people who accept British institutions and ways of governing themselves. (Roger Scruton, quoted in the *Daily Telegraph*, 2 March 1998).

Deconstructing the Countryside Protests

The form and message of the countryside protests were hence the product of a discourse-situated rationalization by the pro-hunting lobby which constructed a 'reality' in which the protests were not about hunting, but rather were an organic uprising of rural people in defence of true British national values of tolerance and freedom. This representation was largely accepted uncritically by the media and politicians, yet it contains contradictions and paradoxes that are open to critique.

Firstly, whilst the conflation of rural values with national values was fundamental to the countryside lobby's discourse, the lobby fairly indiscriminately melded elements of British, English, Welsh, Scottish and Cornish nationalism. Many of the references through which the countryside lobby constructed its claim to represent national values were clearly situated in the English national tradition, and claims for 'British' national values often involved little more than the transposition of 'British' for 'English'. However, the English nationalism that was emphasized was not one that was antagonistic to Scottish, Welsh or Cornish senses of identity. Rather, the positioning of the countryside at the heart of English national identity, and of rural people as the guardians of national values of freedom and tolerance, was echoed by appeals to the same fundamental elements of rurality, tradition and freedom in the very different discourses of Scottish and Welsh nationality.

These appeals to Scottish, Welsh and Cornish nationalism were manifest not only in the carrying of national flags and singing of national songs, but also by the banner carried at the head of the Welsh marches. This bore the bilingual slogan, 'DROS RYDDID COLLASANT EIN GWAED / FOR FREEDOM WE SHED OUR BLOOD', a misquotation from the Welsh national anthem, *Hen Wlad fy Nhadau*. In its original form – 'Dros ryddid gollasant eu gwead' ('For freedom *they* shed *their* blood') – the lyrics, composed by Evan James in 1856, referred explicitly to the mediaeval warriors who fought for the freedom of Wales against the English invaders (Edwards, 1989). Ironically, perhaps, the banner was carried at the head of the combined march into the Countryside Rally in Hyde Park, behind a Scottish Piper and in front of a placard bearing G. K. Chesterton's line: 'We are the people of England, that never have spoken yet'.

What the countryside lobby had hence succeeded in doing – as symbolized in this arrangement – was placing its agenda as being in the common interest of

English, Welsh, Scottish and Cornish national interests. The nationalistic anti-London sentiments of the Welsh and Scottish were combined with anti-metropolitan feelings in rural England. Yet, this manufactured commonality of interest in fact disguised both the antagonistically incompatible aspirations of the various nationalisms, and significant regional and national differences in attitudes towards the policy questions on which the countryside protests were focused. For example, fox-hunting is essentially an English rather than a British tradition. Only twelve fox-hunts operated in Scotland, mostly in the border regions, and the more quintessentially Scottish field sport – deer-stalking – was not affected by the proposed Bill. Similarly, whilst there are more fox-hunts in Wales, many Welsh hunts are foot-hunts, and are therefore imbued with a very different symbolism to their English counterparts, and many are concerned specifically with the function of 'flushing' foxes from dense undergrowth – a practice expressly exempted from the proposed ban. Moreover, by the time of the Countryside March, significant differences had emerged about the relevant importance of the various threats identified. Whilst most marchers from southern England were primarily concerned with defending hunting, many Scottish and Welsh participants appeared to focus more on the economic problems of agriculture.

Secondly, the representation of the countryside protests as an organic uprising of rural people, and the claim of the protesters to be 'the voice of the countryside' may also be critiqued. Conservation groups including the National Trust and the Council for the Protection of Rural England refused to support the protests on the grounds that they were political in nature and concerned essentially with hunting, whilst the Countryside Protection Group challenged the protesters' claim to represent the countryside:

> We want to see the proper countryside issues come to the fore rather than the narrow, sectional interests of the alliance, who have cleverly used the huge financial resources of international hunting groups to present a small minority interest as a major issue. (Countryside Protection Group Spokesperson quoted in *The Guardian*, 27 February 1998, p. 9).

Furthermore, *The Guardian* reported claims that agricultural workers were being coerced into participating in the protests by their employers, with trade unions alleging that threats of dismissal had been made to refuseniks (Beckett, 1998). Noting that many protesters represented 'the most heavily subsidized industry in the country', and that transport had been paid for by landowners and employers, the Chair of the Conservative Anti-Hunt Council commented in a letter to the *Daily Telegraph* that the Countryside March was 'not quite the same as the Jarrow March!' (28 February 1998).

Indeed, whilst the membership of the Countryside Alliance is not homogenous, it is concentrated in particular sections of the rural population. Research by Lusoli and Ward (2003) suggests that the gender, age and income profiles of the 100,000-strong membership are relatively dispersed, but that in occupational terms, farmers and small business owners seem to be over-represented and the rural working class, in particular, is under-represented.

Milbourne (2003) reports that 21 per cent of farmers surveyed in four hunting areas in Cumbria, Exmoor, Leicestershire and Powys were members of the Countryside Alliance, compared with only 5 per cent of non-farmers, and that 12 per cent of 'local' households were members compared with 8 per cent of 'newcomers'. There are also geographical concentrations, as demonstrated by significant differences between Milbourne's study areas. Whilst a third of residents surveyed in the very strong hunting area of Exmoor were members, membership rates dropped to one in twenty residents in Leicestershire and fewer than one in fifty in Cumbria and Powys (Milbourne, 2003). At the same time, the notion of a rural-urban divide is blurred by a reportedly significant Countryside Alliance membership in parts of London, notably Kensington and Chelsea. A MORI poll conducted at the 1998 Countryside March found that 57 per cent of participants interviewed described themselves as living 'in the middle of the countryside' and 22 per cent as living on the edge of the countryside, but 16 per cent lived in towns and cities, and five per cent in the suburbs.

In its senior positions, the Countryside Alliance has continued the precedent set by the Countryside Movement in making appointments that reach out beyond the hunting lobby's traditional allies in the Conservative party. Labour peer Baroness Mallalieu was appointed as the president of the Countryside Alliance in 1998, whilst John Jackson, a founder member of the Country Business Group and chairman of the Countryside Alliance since 1998, is a member of the Fabian Society. In particular, the Alliance regularly highlighted the fact that Richard Burge, its chief executive from 1998 to 2003, was an environmentalist and long-standing Labour party member. These appointments have enabled the organization to counter accusations that it was a Conservative front-group or was pursuing a doggedly anti-government agenda, and to position itself as a non-partisan campaign. Yet, in this way at least the Countryside Alliance board is unrepresentative of its membership. The MORI poll at the Countryside March recorded that 79 per cent of participants had voted Conservative in 1997, and only 7 per cent had voted for Labour. A similar poll at the 'March on the Mound' in Edinburgh in December 2001 found that 83 per cent of participants were Conservative supporters (RPM, 2002).

The aristocratic flavour of the BFSS board was strategically diluted at the formal foundation of the Countryside Alliance in 1998, with the retirement from the board of the Duke of Westminster and the demotion of Lord Kimball, formerly president of the BFSS to deputy president. Nonetheless, the Countryside Alliance board has included several members with roots in the traditional rural land-owning elite, including Lord Mancroft, a Conservative peer; Robert Waley-Cohen, a Midlands businessman and part of an Exmoor landowning family; Sam Butler, grandson of former Chancellor, R.A.B. Butler; and Sir Edward Dashwood, a Buckinghamshire landowner. In September 1999, the *Observer* newspaper, quoting a 'leaked' document, reported that donors to the Countryside Alliance included the Prince of Wales, the Duke of Northumberland, the Duke of Westminster, the Marquess of Hartington and Lord Vestey, as well as three construction companies, Sir Robert McAlpine, Sunley Holdings and Persimmon Homes. Whilst these connections could be read as simply representing the sensible tapping of the resources of notable hunting supporters, they have permitted critics of the Countryside Alliance to accuse the organization of hypocrisy:

> It wasn't easy for the townspeople who watched it go past to work out what the huge Countryside March in March 1998 was all about ... [It] had been called, we were told, to defend the countryside from the town, whose tyrannical and uncomprehending governance of rural areas was leading to the collapse of rural employment and the smothering of farmland by new housing developments. Deferential as ever, we townies were careful not to display our ignorance of rural life by asking who had sacked the agricultural labourers, or who had sold the land to the housebuilders. If the rural environment had been destroyed, rural livelihoods lost and 'country values' dissipated, then, our urban leader writers all agreed, it must be the fault of the cities. (Monbiot, 1998, p. 91).

Thus the positioning of the protests in a radical tradition must be balanced against the use of structures of paternalism, deference and social stratification in their organization. Moreover, the 'reality' constructed for the protests reflected was located in a conservative, exclusionary, discourse of rurality which had been formulated and reproduced by the rural landowning elite over three centuries as a mechanism for legitimizing their position of authority in rural society (see chapter two). This discursive positioning had implications for the fundamental claims made for the protests. The protests were represented as being essentially concerned with defending 'freedom', but that 'freedom' had clear yet undisclosed limits. The protesters demonstrated for the right to hunt, but against the right to roam – the proposed right of public access to open moorland. They opposed state intervention in field sports or in qualifying property-rights, but they demanded state intervention in support of agriculture. Trollope (1998) wrote that the protests were about the 'inalienable right to live in the countryside', yet they objected to plans to build 2.2 million new houses in rural England. The rights and freedoms evoked by the protests existed within the framework of a structured and socially-ordered rural, founded on an implicit convention about the rights and responsibilities of property-owners (see Rose *et al.*, 1976).

Similarly, whilst some speakers at the Countryside Rally, including Baroness Mallalieu and Michael Heseltine claimed the protest to be about defending the British tradition of tolerance towards minorities, other speakers displayed rather less than tolerant attitudes in characterizing their opponents:

> I'm pigsick of weirds with beards of both sexes assaulting me. From now on I will return blow and abuse with compound interest. Stop letting these townie buggers grind us down. (R. F. Poole, quoted in the *Daily Telegraph*, 11 July 1997, p. 1).

In a similar vein, a hunting magazine, *Earth Dog, Running Dog*, described opponents of hunting as 'imposters, actors turned to politics, lesbians and hard women posing as men' (quoted by Vidal, 1998), and was also alleged to have published homophobic and racist comments about Labour politicians.

The countryside protests hence were both rooted in and reproduced an exclusionary discourse of rurality with clear understandings of who may be considered to be truly rural and what it means to be rural. By defining rural people as those adhering to values it claims to represent, the Countryside Alliance defined rurality in relation to itself. In its own 'reality' the Alliance could justify its claim to speak for the countryside. However, by appealing to a ruralist discourse of national identity, and by claiming 'rural' values as national values, the protests mobilized not only an exclusionary discourse of rurality, but also an exclusionary discourse of what it was to be 'British' (or English). Thus Charles Moore used the Countryside March to critique the Labour government's attempts to 're-brand' the image of Britain:

> Taking part in yesterday's march, I was struck even more forcibly than before with the utter absurdity of Cool Britannia. Among the 280,000 or so smiling faces, with caps above them and tweed below, I could see not one single person who answered the Mandelson depiction of our nation. Warm, yes; cool, no. (Moore, 1998, p. 20).

The positioning of the rural at the heart of the construction of national identity hence has consequences beyond the defence of rural interests, and the countryside protests in mobilizing this discourse cannot be read simply as demonstrations about hunting or rural services or agricultural problems, nor even about freedom and democracy, but must also be read as contributions to a wider debate about British national identity at the close of the twentieth century.

Internal Disputes and Militant Splinter Groups

The success of the Countryside Rally and the Countryside March fuelled a debate within the hunting lobby as the merger between the BFSS, Countryside Movement and Country Business Group to form the Countryside Alliance was formalized in the spring of 1998. For the core hunting supporters and the established leadership of the BFSS, the broadening of the agenda to embrace wider 'rural' issues had always been understood to be a ploy, and there was concern that the formal constitution of the Countryside Alliance marked a diminishing of the focus on hunting as the prime issue. A second group, with origins in the Country Business Group and particularly associated with Eric Bettelheim, adopted a more business-oriented perspective, understanding the Countryside Alliance as a brand that needed to be sustained as the media focus passed on from hunting. Essentially, that meant identifying new 'markets' and the development of the Alliance's campaigning in areas other than hunting appeared to be a way to achieve this. The strategy was openly ridiculed by more hard-line hunt supporters, including the Alliance's outspoken press officer, Janet George:

> The decision to allow ramblers and canoeists to join annoyed many of the members – ramblers are only of use when we run out of pheasants to shoot, and canoeists scare off the fish. The Countryside Alliance can't be all things for all people. (Janet George, quoted in Carter, 1998, p. 3).

The internal dispute in the Countryside Alliance crystallized around a personal feud between Janet George and Edward Duke, a Yorkshire businessman who took over as chief executive following the Countryside March (see George, 1999, for one perspective). George resigned in June 1998, but her popular following among grassroots hunting activists and with the field sports press produced a reaction that further destabilized the Countryside Alliance and led to resignations of both Duke and the Alliance chairman, Charles Goodson-Wickes, a former Conservative MP. Ironically, the subsequent appointment as chairman of John Jackson, chairman of Mishcon de Reya law firm, and as chief executive of Richard Burge, arguably strengthened the position of those who sought to expand the organization's profile. Richard Burge, in particular, reportedly provoked concerns with an interview to the *Spectator* magazine 'in which he said that he would take a job from anybody, had never hunted, and that the Countryside Alliance would survive without hunting' (RPM, 2002, p. 46).

Under the leadership of Jackson and Burge the Countryside Alliance has established itself as a major pressure group with a considerable presence in the media and in policy debates. Yet, it has experienced both criticism from some quarters in the hunting community for building too many links with the government, and a growing sense of discontent among its membership about its ability to ultimately prevent the passage of anti-hunting legislation. These concerns have become manifest in an increasing militancy among grassroots hunting activists. The prospect of adopting more confrontational tactics had been hinted at for some time. Maurice Askew, a former member of the Countryside Alliance board and leading member of the Union of Country Sports Workers, reportedly told the Welsh Countryside Rally in Cardiff in November 1999 that, 'My warning for Mr Blair is that if he does not listen to us there will be a civil war in this country, the like we have never seen since the days of Cromwell and Fairfax' (*Western Mail*, 11 November 1999), whilst the first obstructive direct action was undertaken in July 2000 with a blockade of the Severn Bridge by the Rural Action Group. At the time, however, such warnings and activities were considered to belong to the fringe of the countryside movement.

By the Liberty and Livelihood March in September 2002 a wider disposition towards militancy had begun to emerge. The optimism that had pervaded the earlier mass demonstrations was replaced in the chatter of participants with a sense of frustration and defiance. The overheard comment of one marcher that, 'next time we'll come back with tractors' was typical of the sentiments expressed, as was the placard that declared 'This is our last peaceful march' (Figure 5.2). Participants in the march were also encouraged to sign up to the Hunting Declaration, an online initiative sponsored by Sam Butler and the philosopher, Roger Scruton, through which individuals pledged to break the law by continuing to hunt if the sport was banned. The statement had gathered some 37,000 signatures by 'Declaration Day' on 1 November 2003, with a further 19,000 names added in the five months up to April 2004. Similarly, a poll by the Farmers Union of Wales found that over 60 per cent of hunt supporters questioned at the Royal Welsh Winter Fair in December 2003 would be prepared to break the law if hunting with hounds was made illegal (Brecon and Radnor Express, 15

January 2004). Against this background of grassroots militancy, the Countryside Alliance itself has hardened its position. In a notable article in the Guardian in October 2002, the Alliance's chairman, John Jackson, drew on the work of Henry David Thoreau to argue that whilst the 'Countryside Alliance would never recommend its members, or others, to engage in civil disobedience', it would be prepared to support those who, as a matter of individual conscience, elected to do so:

> If there were a ban, anybody – whether or not a member of the alliance – who regarded that ban as manifestly unjust and, as a matter of conscience, followed Thoreau's thinking would be supported publicly by the alliance, particularly at the time of their trial. (Jackson, 2002, p. 20)

Figure 5.2 Placard at the Liberty and Livelihood March

A more militant strand of rural activists, however, have become increasingly critical of the Countryside Alliance, arguing the need for radical direct action before, not after, a ban on hunting. In particular, these activists have looked to the tactics of direct action employed by environmental campaigners and by militant farm protests, including the fuel blockades of September 2000 (see chapter six), for inspiration, and have organized themselves into a number of radical splinter groups. The largest of these is the Countryside Action Network, founded by the former Countryside Alliance press officer Janet George, which claims to have 4,000 members. Whilst the Countryside Action Network can be argued to be the most cautious of the breakaway groups, it has advocated civil disobedience and non-violent direct action, organizing road blockades and go-slow rolling motorway protests. George, who distinguishes her organization from the Countryside Alliance by asserting that 'they are a political lobby group – we are an action group' (BBC News Online, 12 July 2002), describes the hardening of tactics as part of an escalating conflict between the countryside and the government:

> People are starting to realise we have a war on. It's really a case of doing enough to put the wind up the Government without alienating the general public too much. We need to raise the stakes in terms of irritating and inconveniencing the Government. (*The Observer*, 22 December 2002).

Similar tactics have been adopted by the Rural Rebels, a splinter group based in Scotland and whose actions have been primarily focused on the Scottish Parliament. Claiming a membership of around 100, but support from 19 countryside protest groups, the Rural Rebels campaign on a three-fold agenda that positions the right to pursue country sports alongside demands for an inquiry into the Foot and Mouth epidemic and an end to the closure of rural schools. Wearing orange boiler-suits as a highly visible mark of identification, members of the Rural Rebels have mounted 'rolling blockades' of tractors and slow-moving vehicles on the M74 motorway, Forth Road Bridge and in central Edinburgh. Like other rural splinter groups they attribute their militancy both to the need to confront urban interference and to the failure of conventional political methods.

The most militant and most controversial of the breakaway groups, however, is the Real Countryside Alliance, or Real CA, launched in May 2002, whose name and organizational structure are modeled on that of the Real IRA. The Real CA claims to have around 800 members organized into 100 'cells' that are able to operate autonomously. It is also claimed to have significant financial support, which enabled it to launch itself with a billboard poster campaign juxtaposing the figure of a terrorist with the caption 'freed', with that of huntsman with the caption 'imprisoned?'. Arguably, the Real CA has been most faithful to the principles of 'post-modern protest' in its activities, largely focusing to date on symbolic stunts aimed at capturing media attention, including adding giant huntsmen to historic chalk 'white horses' in Yorkshire and Oxfordshire, and hanging a banner proclaiming 'Love Hunting' from the Angel of the North on Valentine's Day 2003. Stickers with the group's green union jack logo (Figure 5.3) have also been attached to road signs in rural areas, serving as territorial marks, as well as being used to deface the offices of Labour MPs. One such sticker carried the warning, 'Civil Rights or Civil War'. These actions in themselves have technically involved criminal damage, and the Real CA has boasted of 'escalating criminal activity' in resistance to anti-hunting legislation. Edward Duke, the former chief executive of the Countryside Alliance who is 'associated' with the Real CA, informed the *Daily Telegraph* that:

> Everything so far has been graded by the Real CA as between levels one and three in terms of extremism. One cell leader was discussing how to saw the leg off an electricity pylon. Gas and water supplies will also be targeted and lorries carrying food for supermarkets. (Edward Duke, quoted in the *Daily Telegraph*, 17 November 2002)

Figure 5.3 Green Union Jack logo of the Real Countryside Alliance, attached to a rural road sign

A similar threat was conveyed in an article published in *The Field* magazine just before the Liberty and Livelihood March in September 2003, which warned:

> When the country-wide hunt supporters are mobilised they can do so much more than tramp the streets of London. For example, the general public will wake up one morning and hear there is not a single speed camera working in Britain. Their reaction will be, 'Say what you like about the hunting boys, a least they've got a sense of humour.' Ministers will react differently, and mutter, 'These people are well-organised. They're not irresponsible, and if they really decide to turn the screws, God alone knows what they'll do.' It may at a later date be necessary to alienate everyone, but please God that will not be necessary. Suppose someone were to pull the plug out of a reservoir in Wales and run Birmingham short of water; suppose every motorway were blocked. Anger breeds chaos. (Walton, 2002, p. 52).

To date, this rhetoric has not been matched by action. Yet, in giving voice to such ideas, militant rural protesters have employed the language of intimidation as a weapon in its own right, with the intention of threatening the government, pressurizing the Countryside Alliance and frightening the public. It also serves to reproduce the idea of the 'true rural population' as a beleaguered, besieged, community, reinforcing the perceived rural-urban divide and encouraging rural solidarity. However, the discourse of powerlessness and oppression still sits uneasily with the resources that are available to the militant rural groups and with the pedigree of some of their most prominent supporters. Leaders of the groups include successful businesspeople, hunt masters, large farmers and landowners, many of whom are part of the traditional rural landowning elite. Hence, the rural movement in Britain can be argued to have attempted to follow the reverse of the normal trajectory of social movements, moving not from an outsider status to an insider status, but from insider status to outsider status. Having historically relied on informal insider networking and lobbying to protect the interests of hunting, the message from the militants is that hunting can now only be defended from the outside, by engaging in disruptive, sometimes illegal, actions. The Countryside Alliance itself may be described as a 'thresholder group', that vacillates between 'pursuing and not pursuing a symbiotic relationship with decision makers' (May and Nugent, 1982). As such, it exhibits 'strategic ambiguity and oscillation between insider and outsider strategies' (*ibid.*), the former including political lobbying, involvement in policy consultations and exercising influence within sympathetic elite fractions, including the parliamentary Conservative party and the House of Lords; and the latter involving 'appeals to the public through the mass media and efforts at the broad scale mobilisation of citizens as "grass-roots"' (Walker, 1991).

Whatever public antagonism is expressed between the Countryside Alliance and the militant protest groups, the two blocs are ultimately mutually dependent. The radical protest groups need the Countryside Alliance because the Alliance's engagement with policy makers is the only avenue through which the protesters' demands can be followed up and desired outcomes agreed, and because the presence of the Countryside Alliance means that groups such as the Countryside Action Network and the Real CA get treated as part of a mass movement, not as an extremist fringe. At the same time, the Countryside Alliance needs the radical protest groups for the media coverage and pressure on politicians that they are able to generate, and to act as a 'pressure valve' for their own members. Indeed, this complex web of inter-dependence extends beyond the pro-hunting groups discussed here to a parallel relationship between established farming unions and militant farmer protest groups, as examined in the next chapter.

Chapter 6

Agricultural Politics

Agriculture and Rural Politics

Rural politics in Britain were for a long time largely conflated with agricultural politics. Grant (1990), in a review of research on rural politics in Britain, noted that studies had tended to focus primarily on the politics of agriculture, yet this academic bias essentially reflected policy and popular assumptions. The discourse of the agrarian countryside, discussed in chapters two and four, which came to prominence in the mid twentieth century, reinforced the identification of the rural economy and society with farming. It functioned not only as a discourse of rurality but also a discourse of power, setting the parameters of political discussion and policy-making on rural issues, and positioning farmers' leaders as the appropriate political leaders and representatives of rural areas. The dominance of the agricultural elite in rural local government (see chapter two), was accompanied by growing agrarian influence in the Conservative Party (see chapter three), giving the farming lobby a national voice that was institutionalized as the National Farmers' Union and, to a lesser extent, the Country Landowners' Association, were incorporated into policy-making mechanisms. To some extent, the priority given to agriculture in these developments accurately reflected the position of farming as a major economic activity and employer in the countryside, but even by the 1950s little more than a third of the rural population were dependent on agriculture and the claim of the agricultural lobby to speak for the countryside was beginning to be challenged:

> On so many occasions the farmers as a body seem to be against everybody else, for instance with regard to prices, wages, town and country planning, school education etc. On these subjects farmers usually claim that they speak for the countryside. They do – the farming countryside as seen through the eyes of an agricultural employer. Their wider claims to speak for all countrymen may be regarded as arrogance by some of the non-farming members of the rural community. (Bracey, 1959, p. 45).

The formal incorporation of the farming unions into the policy process started during the First World War as the government sought to secure food supply by introducing measures including guaranteed prices for wheat, guaranteed wages for farm workers, and powers over the cultivation and stocking of farmland (Winter, 1996a). These measures were implemented through county agricultural executive committees that directly involved representatives of the NFU in agricultural policy for the first time. Although the inter-war period saw a lessening of state regulation of agriculture, the system of agricultural committees was maintained in a revised form. County committees elected representatives to national councils of agriculture, which in turn provided an input to a smaller Agricultural Advisory Committee that worked closely with the Ministry of Agriculture. The NFU, whose membership more than doubled between 1916 and the early 1920s, successfully secured representation at all levels, including five guaranteed places on the Agricultural Advisory Committee (Winter, 1996a).

It was the Second World War, however, that provided the impetus for a new agricultural settlement in Britain based on significant state intervention in agriculture and the direct involvement of the NFU in policy-making and implementation. Emergency powers were again employed to regulate food production under the direction of war agricultural executive committees, most of which were chaired by leading farmers or landowners who were active in either the NFU or the CLA. Moreover, in contrast to the post-First World War period, there was also wide recognition of the need for continued state intervention after the war had ended. In 1944, the NFU, CLA and nine other organizations including the councils of agriculture for England and Wales and the two main trades unions representing farmworkers combined with a group of peers to set out a plan for agricultural policy that emphasized the need for increased food production assisted by price support, subsidies for land improvement and credit facilities for farmers (Winter, 1996a). These elements were subsequently incorporated into the Agriculture Act 1947 which not only established the post-war agricultural policy, but also when taken together with the Town and Country Planning Act 1947 and the National Parks and Access to the Countryside Act 1949 formed the framework for an approach to rural policy in Britain in which the interests of agriculture were implicitly prioritized by the very act of policy segmentation.

The Agricultural Policy Community

The high level involvement of the NFU in the agricultural policy process in the period following the Second World War has been described as 'the paradigm case of a closed policy community' (Smith, 1993, p. 101). Policy communities, as conceptualized by Marsh and Rhodes (1992), are the most tightly-constituted type of policy network or structure through which a range of actors are engaged in the policy process. As such they are defined by a limited number of participants, the predominance of economic and professional interests, frequent and high-quality interaction on all matters related to policy issues, continuity of membership and consistency of values and outcomes, an exchange of resources between participants with the leaders of participating organizations delivering results to members, and a balance of power between members (Marsh and Rhodes, 1992). All of these characteristics were present in the agricultural policy community, as Winter (1996a) details.

Firstly, the participants shared an economic interest in increasing agricultural production – identified as being of importance to the state because of the need to ensure an adequate food supply and to maintain agriculture's contribution to national economic performance, as well as to farmers themselves for reasons of self-interest. Secondly, there was consensus among policy community participants over the need for public investment and expenditure to support agriculture. As Winter (1996a) notes, the backing of the NFU was harnessed by the Ministry of Agriculture, Fisheries and Food (MAFF) in internal negotiations for government resource. Thirdly, there was agreement between participants that securing food supply was the primary problem for the rural economy and that state intervention and investment was the only way to achieve this. Fourthly, the participants shared a 'common culture' that took as granted a belief in agricultural progress through scientific innovation, in the application of management economics to improve farm efficiency, in agriculture's import-saving role, in the need to protect farmland from development, in opposition to land nationalization, in the environmental benefits of economically sound agriculture, and in the moral value of farming and its central role in both rural and national life (Winter, 1996a).

Most significantly, the agricultural policy community was characterized by the exclusivity and stability of its membership. The inner circle of the policy community, which was engaged in day-to-day policy making and strategic planning across all areas of agricultural policy, consisted only of officials of MAFF and the NFU. It was at this level that the shared culture and consensus of opinion was most tightly maintained, even to the extent of excluding civil servants who dared to question the principles underlying agricultural policy (Smith, 1989). Beyond the inner circle, an outer circle of the policy community included organizations such as the CLA, the National Union of Agricultural and Allied Workers, and major food processors, whose involvement was largely restricted to specialist areas. Excluded altogether were those groups most likely to dissent with the principles of the policy, notably environmental groups, animal welfare groups and consumer groups. The consequence was to create an illusion of unanimity in agricultural policy-making and to justify the continuing stability of the policy community's membership:

> By excluding groups which disagree with agricultural policy, the community is able to say that there is a consensus on agricultural policy. Consequently the consensus demonstrates that there is only one possible agricultural policy, and hence that there is no need for consumer representation, as the community ensures that agricultural policy is in their interests. (Smith, 1992, p. 32).

Particularly striking is the way in which the policy community not only excluded non-agricultural interests, but also created a two-tier hierarchy within the agricultural sector by privileging the position of the NFU. As Winter (1996a) describes, the union's membership of the inner circle helped it to secure a virtual monopoly of agricultural representation in England and Wales. Organizations in the outer circle of the policy community were content for the NFU to take the lead in return for influence in their own specialist area. Even the CLA, which had initially been the NFU's great rival, had found an accommodation, leaving the NFU to lead negotiations in areas such as commodity issues whilst developing its own specialisms in fields such as property law and public access, where the NFU was less active. The NFU hence became a classic example of an insider interest group with a hierarchical structure of local and county branches and commodity committees serviced by a cadre of professional officers. Membership of the NFU became 'almost compulsory' for farmers (Winter, 1996a), not only because of its lobbying activities but more pragmatically because of its legal and insurance services, its provision of technical advice and its involvement in marketing and other commercial ventures. The mass membership of the NFU – which peaked at 210,000 in the early 1950s and included 90 per cent of farmers in England and Wales by 1957 – served as a bargaining tool to reinforce the NFU's position in the backstage lobbying that formed the core dimension of the union's political work. Notwithstanding its support in the parliamentary Conservative Party, the NFU preferred to ensure that agricultural legislation was 'right' before it reached parliament (Grant, 2000), and its relationship with MAFF was sufficiently close that the Ministry became in effect the NFU's advocate *within* government. NFU members were not, therefore, required to mobilize politically in any direct or overt way, and any temptation to do so could be contained as long as the NFU leadership was perceived to be delivering satisfactory results to grassroots farmers.

On the odd occasions when challenges were launched to the NFU from within the farming community, the union tended to react by seeking to exclude the dissident bodies from the policy community and thus neutralizing their influence. The most notable case of this is the Farmers' Union of Wales (FUW), which was formed in 1955 by upland sheep farmers and marginal dairy farmers who perceived the NFU to be dominated by lowland, English, arable farmers and thus unsympathetic to their interests (Murdoch, 1992). The membership of the FUW was drawn mainly from Welsh speaking farmers in west and north Wales and increased with the rise of Welsh nationalism in the 1960s and 1970s. Throughout this period, however, the FUW was faced with hostility from the NFU and excluded from the policy community. It only gained full consultative status from MAFF in 1978 as a result of lobbying by Welsh Liberal MP Geraint Howells during the Lib-Lab pact.

Britain's accession to the European Economic Community in 1973, subsuming of British agricultural policy within the Common Agricultural Policy (CAP), did little to change the exclusivity of the policy community or the dominance of the NFU. The policy community continued to control the implementation of the CAP in Britain and to determine the British position in EEC negotiations on agriculture. Furthermore, the NFU enjoyed a presence in the European agricultural policy community through its membership of COPA (*Comité des Organizations Professionelles Agricoles*), the Europe-wide federation of the major farm unions. COPA had been heavily involved in devising the CAP and continued to enjoy a close and closed relationship with DG6, the department of the European Commission responsible for agriculture (Grant, 2000; Winter, 1996a). Arguably, the NFU took advantage of EEC membership to enhance its position as the primary representative of British advocate, blocking FUW membership of COPA and being prepared to directly lobby the European

Commission on occasions when the position adopted by COPA was considered to be unfavourable to British farmers.

Collectively the policy communities in Whitehall and Brussels oversaw the pursuit of productivism in British agriculture in the three decades after the Second World War, a strategy driven by the objectives of maximizing agricultural production and maintaining the standard of living for farmers. This approach produced major changes in the practice of farming in Britain. The intensification of agriculture sought higher productivity through substantial capitalization, including significant investment in machinery and the farm infrastructure and the utilization of agri-chemicals and other biotechnologies. The concentration of agriculture aimed to maximize cost-efficiency by creating larger farm units. Finally, the specialization of agriculture also helped cost effectiveness, with farm units concentrating on a single produce, often sold under contract to a single purchaser (Woods, 2005a). In its own terms, productivism was highly successful. Not only did total agricultural production increase significantly, but the basic structure of the British family farm was maintained. As Gilg (1984) summarized, the post-war settlement produced,

> an agriculture which provides a high proportion of Britain's dietary requirements at a reasonable price, many villages have been attractively developed or conserved, large areas of open countryside remain scenically beautiful, much bad development has been prevented and much pollution of the air and water has been cleared up. (Gilg, 1984, p. 250).

However, the exclusive nature of the policy community meant that it turned a blind eye to the negative consequences of productivism. Changes in farming practice, for instance, had a significant detrimental impact on the environment, resulting in the loss of hedgerows, ponds and meadows (both as landscape features and as habitats for wildlife), pollution and poisoning from agri-chemicals, dramatic reductions in the population of farmland birds and animals and of several native plant species, and problems of soil erosion and flooding (Harvey 1998; Woods, 2005a). Factory farming crowded farm animals into inhumane conditions, and concerns grew that the move towards mass, standardized food production came at the expense of food quality – a concern that was later manifested in a series of food scares. Furthermore, productivism was arguably too successful in its core objective of increasing food production. By the 1970s, European agriculture was producing greater quantities of key commodities than could be sold at a reasonable price on the open market. In order to maintain prices – the other core objective of the CAP – the European Commission was forced to intervene and buy surplus produce once a minimum price threshold was reached. Thus, in 1982 the EC held nearly seven million tonnes of surplus common wheat stocks in storage, with similar (though comparatively smaller) stockpiles of surplus beef carcasses, butter, milk and barley (Winter, 1996a). As well as the cost of buying surplus stock, the cost of storage became a major financial burden on the EC, increasing by nearly 400 per cent between 1973 and 1984. Overall, the implementation of the CAP accounted for 70 per cent of the entire European Commission budget in 1984, with a quarter of that spent solely on storing surplus stocks (Woods, 2005a). In such circumstances, the rationale of the CAP could no longer be justified in terms of securing food supply. Instead, the CAP became in effect an 'agricultural welfare state' (Sheingate, 2001), existing primarily to support the livelihoods of farmers through state handouts and protecting their businesses from the uncertainties of the market.

As such, by the 1980s the CAP was under attack from all sides – from free market advocates on the right who objected to the scale of state intervention and subsidization (see Body, 1982); from environmentalists, animal welfare campaigners and consumer advocates; and from 'progressive' farmers who were experimenting in areas such as organic agriculture. These cumulative pressures helped to force the slow and contentious process of reforming the CAP from the mid 1980s onwards, with measures introduced that aimed to control or reverse the trend of increasing production in over-provided commodity sectors and to reverse some of the worst excesses of productivism. Among the strategies adopted as part of this 'post-productivist transition' were quotas for milk production, payments to farmers to 'set-aside' land from production for a specified period, and grants to assist farm diversification and agri-environmental schemes (Ilbery and Bowler, 1998; Woods, 2005a).

The very fact of CAP reform was a sign that the strength and exclusivity of the agricultural policy community was weakening and that the voices of groups other than the farming unions were being heard. At the same time, in order to maintain their position within the policy-making framework, farm unions needed to engage with the new agenda, which often meant accepting compromises that were not wholly favourable for all or some of their members. As such internal tensions developed in the agricultural lobby. COPA, for example, found itself sometimes forced by its members to adopt 'extreme positions which do not take account of the changed character of the policy debate' (Grant, 2000, p. 97), and other times party to reforms that had been resisted by some of its member organizations. Thus, COPA became increasingly fractious in its own internal politics, and found itself losing influence in EC agricultural policy-making:

> Beginning in the mid-1970s, COPA lost cohesion and by the 1980s was no longer at the centre of the policy process. When the Commission dropped the objective method and began to address more divisive issues, COPA faced dissension within its ranks as individual commodity groups and national organizations clashed over demands from the Community. The needs of individual member of COPA diverged and in some cases conflicted. The national farm organizations moved into the vacuum with greater consultation with their national governments in an effort to maintain control of the farm policy. (Phillips, 1990, p. 73).

Domestically, the NFU faced a similar dilemma. The establishment of the Food Standards Agency in 1999 – after two years of discussions – is one example of the changed position of the union. The NFU initially opposed the idea of the agency, but 'weakened in the face of public demand ... it has concluded that, on food safety issues, MAFF is no longer being taken seriously' (*Farmers Weekly*, 3 January 1997, quoted by Grant, 2000, p. 71). As Grant describes, not only had consumer interests taken precedence over producer interests in shaping political opinion on the need for a Food Standards Agency, but the NFU discovered when it canvassed farmer opinion that its members were almost evenly split on the form that any new agency should take. Indeed, the increasing diversity of opinion within the agricultural sector and the differential impact of reforms on different types of farm is one of the main challenges faced by the NFU. As a perception has developed that the NFU is no longer delivering results to grassroots farmers, so farmers have increasingly contemplated more militant forms of direct action, particularly as farming entered a series of inter-connected 'crises' beginning with the BSE scare in 1996.

BSE: An Agricultural Crisis

Bovine spongiform encephalopathy (BSE) was a disease created by the excesses of productivist agriculture. A brain disorder which also became known as 'mad cow disease', BSE was first detected in a cow on a farm in Surrey in 1983, although the first officially confirmed case was not recorded until 1986 (Woods, 1998d). BSE was a new disease in cattle, but other conditions of a similar type, transmissible spongiform encephalopathies (TSEs), existed in other animals, most notably scrapie which had been recorded in sheep for at least 200 years. The transmission of scrapie-contaminated prions from sheep to cattle in offal from infected sheep that had been fed to cattle as cheap feed, and the related practice of feeding rendered cattle offal back to cattle, were identified as the sources of the new disease, which spread with epidemic force through the British cattle population. Between November 1986 and September 1998, a total of 172,438 cases of BSE were reported in Britain, infecting 37 per cent of all British cattle herds and 61 per cent of all dairy herds. Significantly, the epidemic was largely confined to Britain. Fewer than two thousand cases were reported in fourteen other countries up to the end of the century, although the discovery of BSE in German cattle provoked a political crisis that prompted the resignation of the Agriculture Minister, and suspected cases in Canada and the United States in 2003 caused widespread concern. The introduction of a ban on the use of offal in animal feed from 1988 successfully curtailed the spread of the disease, once the incubation period was taken into account. New cases of BSE peaked in 1992, with 36,772 cases confirmed that year, but by 1995 incidence had fallen to around 10,000 new cases (Taylor, 1996; Woods, 1998d).

However, if BSE could be caused by the transmission of contaminated prions in feed from sheep to cattle, could the consumption of BSE-infected meat cause disease in humans? There was no record of humans being harmed by eating scrapie-infected lamb or mutton and thus the Ministry of Agriculture, Fisheries and Food (MAFF) initially insisted that BSE posed no threat to human health. Yet, the BSE epidemic coincided with a growth in incidences of a new variant of the human TSE, Creutzfeldt-Jakob disease (CJD). CJD had been a long established but rare disorder that had only affected around two million people in total before the 1980s (Macnaghten and Urry, 1998), usually in old age. The new variant, in contrast, was reported in unprecedented numbers in young people, with another unexplained concentration in people associated with farming (Ratzen, 1998).

In March 1996, scientists at the National CJD Surveillance Unit in Edinburgh published their conclusion that the development of new variant CJD as a result of exposure to the BSE agent was 'the most plausible interpretation' (Will et al., 1996, p. 925). The response was immediate and dramatic. Consumer demand for beef collapsed by up to 30 per cent (although it later stabilized at 94 per cent of the previous level), and the European Commission introduced an immediate ban on the export of cattle, beef or bovine products from the United Kingdom (Woods, 1998d). North (2001) suggests that reaction to the acknowledgement of the BSE-vCJD link moved through the five stages of a classic 'scare'. Firstly, in the 'public phase' the story was picked up and given extensive coverage by the media, shaping public opinion. In the second, 'reaction phase', consumers responded by choosing not to buy beef. Although the extreme consumer reaction may be short-lived, a scare is pro-longed by the emergence of a third, 'political phase', in which politicians seek to frame a response. This leads in turn to a 'legislative phase' in which new legislation or regulations are introduced, and finally to the 'enforcement phase' in which the new regulations are enforced, usually long after public concern has subsided. The BSE scare followed this model, but it was more than just a food scare. As Macnaghten and Urry describe, the BSE problem escalated into a multi-faceted political crisis for the British government:

> The BSE crisis has had profound political and economic consequences. These include, inter alia, the effective collapse of a £500 million beef export industry, a longer-term ban on British beef across the European Union, a number of rejections by EU states of UK proposals to ease the beef ban, the slaughter of over a million cattle ... and rising costs of measures to deal with BSE, estimated in late 1996 at over £2.5 billion. For a short time the British Government even set up a UK task force to veto all EU decisions until the EU agreed to a clear framework to ease the ban. What to the Government started as a purely technical problem of how to deal with a new disease in cattle escalated into what PM John Major described as 'the worst crisis the Government had endured since the Falklands'. (Macnaghten and Urry, 1998, pp. 258-9).

The key challenge facing the government was to mitigate the economic impact on agriculture by seeking to restore consumer confidence in British beef and securing the lifting of the export ban. It thus introduced a wide-ranging eradication programme, including a cull of all cattle over thirty months old. Over 2.6 million cattle were slaughtered in the cull, including more than one million in the first six months of the scheme. The impact of cull, however, was spatially uneven, reflecting the uneven geography of the BSE epidemic. Around 45 per cent of all cases of BSE in Britain had been recorded in just eight counties – Devon, Somerset, Cornwall, Dorset, Dyfed, Cheshire, North Yorkshire and Wiltshire – and similarly, Devon, Dorset, Cornwall and Somerset accounted for a quarter of all cattle registered from the 'Over Thirty Months Scheme' in November 1998 (Woods, 1998d). Similar geographical patterns were evident in the overall economic effect of the crisis. The official MAFF report in 1998 suggested that only around 1,000 jobs had been lost as a direct result of the crisis in the first twelve months (compared to projections of up to 46,000 job losses), and that the slump in beef sales had been compensated for by increased demand for poultry, lamb and pork. Up to 500 new jobs had been created in food regulation. Yet, this aggregate balance sheet disguised the geographical winners and losers:

> The biggest losers were in Northern Ireland, followed by Scotland, and parts of northern and south-west England. In eastern counties of England and lowland areas, where pig and poultry farming are important, farmers gained, while their counterparts in the upland and western areas lost out. (*The Guardian*, 14 March 1998).

Ultimately the programme was successful in its objectives of eradicating BSE, restoring sales of beef and securing the end of the export ban, which was lifted in November 1998. The financial cost, however, was estimated to be between £740 million and £980 million, and the political cost was growing tension between farmers and the government.

Farmers' dissatisfaction with the government's handling of the BSE crisis was rooted in three factors. Firstly, many farmers felt that they had been *betrayed* by the government which had failed to challenge connection between beef and vCJD. Agricultural lobbyists pointed to the imprecision of the scientific judgement and the qualification in the paper that sparked the crisis that 'we emphasise that we do not have direct evidence of such a link [between consumption of BSE infected beef and vCJD] and other explanations are possible' (Will et al., 1996, p. 925); and the NFU ran a publicity campaign claiming that 'British beef is safe' and 'there is no scientific proof that BSE can be transmitted to man by beef' (see Woods, 1998d).

Secondly, farmers were *cynical* about the motives behind the European Commission's ban on British beef exports, and in particular about France's refusal to permit imports of British beef long after the EU ban had been lifted. The ability of other EU states to continue to export beef *to* Britain during this period was perceived as deeply unfair and became a focal point for irritation. Thirdly, some farmers considered that the government had *unnecessarily intensified* the impact of the crisis by insisting on the slaughter of healthy cattle as a precautionary measure. Opposition to this policy was articulated in highly emotional language that revealed the close attachment felt by many farmers for their stock (see also Woods, 1998d):

> Working as a dairy and beef farmer seems the hardest way to earn a living. These farmers regard their cattle as an extended family – they have known nothing else. Every animal is a friend, regardless of the endless hard work entailed. (Letter to *Farmers Weekly*, 24 May 1996).

> All I do is beef. I have't got a very big farm and I'm too close to the town for sheep. For me it's beef or nothing. I hope it won't come to slaughter, I'd cry. (Pembrokeshire farmer quoted in the *Western Mail*, 22 March 1996).

I'm going back to a life of getting up at 3am to help with calving just to produce bull calves for slaughter. (Somerset farmer quoted in *Farming News*, 11 October 1996).

The animal rights lobby has made its beliefs known to the farming community and, over many years, has caused disruption, damage and injury to the business and its workers. For no logical reason, farmers are now forced by the EU to slaughter hundreds of thousands of healthy bovines and I have not heard or seen any sign of protest from the activists. (Letter to *Farmers Weekly*, 17 May 1996).

Plaintive though the appeals of farmers were, they were also out of line with the new prevailing political climate on agriculture. Farmers saw themselves as the victims, but public opinion blamed farmers for the problem of BSE, and by extension, for a threat to human health. As Whatmore (2002) observes,

The ethical (and political) import of the BSE-vCJD epidemic in Britain begins by acknowledging the corporeal specificities of cows as herbivourous ruminants, and following the incongruous 'rationale' of a feeding regime indifferent to them through to the eating habits and food choices of consumers. The practice provoked revulsion and disbelief in equal measure among an unsuspecting public. What kind of rationality was it that could make sense of such routine cannibalism? (Whatmore, 2002, p. 163).

Furthermore, the revelation reinforced other, broader, concerns about the impact of productivist agriculture on the rural environment and contributed to a loss of public confidence in farmers' stewardship of the countryside – the idea that had been at the heart of the discourse of the agrarian countryside:

the popular image [of farming] took something of a battering following the BSE crisis. The revelation that dairy farmers had fed cattle carcasses back to cattle came as a shock to many. There were had been other concerns too – the continuing destruction of hedgerows, the calf trucks which, until the European beef ban, lined the dockside, the seemingly constant recourse to pesticides. (Harvey, 1998, p. 2).

This shift in public opinion gave the upper hand in political debate to advocates of agricultural reform and to environmental and food safety campaigners, whilst isolating the position of the farm unions to the apparent detriment of the ordinary farmer:

The NFU's response to the latest BSE scare has again alienated the British farmer from the public. The farmer and his beloved cows are the losers in this fiasco. (Letter to *Farmers Weekly*, 5 April 1996).

As the NFU was increasingly perceived as being unable to deliver satisfactory results to the farmer, so farmers began to contemplate more radical forms of resistance. One FUW leader suggested that 'our farmers are willing to protect their animals by getting guns if necessary ... Welsh farmers are going to defend their farms' (quoted by the *Western Mail*, 18 April 1996), a refrain of militancy that became more frequent as problems spread through the various sectors of farming from 1997 onwards.

The Manufacture of a 'Rural Crisis'

The use of the word 'crisis' to describe any of the challenges that have confronted British farming since the mid 1990s is in itself controversial. Forbes (2004), for instance, has disputed the description of the events surrounding the acknowledgement of a link between BSE and vCJD – discussed in the last section – as a 'crisis'. It was, he argues, not a crisis but a *drama*. The distinction is useful: Forbes suggests that refer to a 'crisis' is to prematurely invoke a political judgement about what happened, and in the case of BSE Forbes concludes that the implications of the word 'crisis' are not justified by more careful analysis. In contrast, referring to a 'drama' emphasizes the *manufacture* and *performance* of the narrative of events. For the BSE affair, this includes the ways in which a very precise, technical, finding of a piece of scientific research was made intelligible to a wider audience such that it initiated significant political and economic effects. Most commentaries on the performance of the BSE affair have focused on the use of scientific knowledge as part of this process (see Hinchcliffe, 2001; Jasanoff, 1997; Leach, 1998; Miller, 1999; Whatmore, 2002; Winter, 1996b), however, in understanding the political implications for the countryside, the representation of the BSE affair as a *rural crisis* is equally significant.

The recasting of the problems faced by beef farmers as a consequence of the export ban, fall in consumer confidence and cull of cattle over thirty months old, as a 'crisis' for the countryside as a whole, was part of a discursive strategy by farmers to regain public sympathy. The performance of this representation involved three elements. Firstly, the historical position of farming as the cornerstone of the rural economy was invoked to suggest that the financial problems faced by individual farmers could destabilize the whole rural economy and, by extension, rural society:

> It's drastic action we want now and we need it immediately. Otherwise this crisis will cripple not only the beef industry, but the whole rural community. (Pembrokeshire farmer quoted in *The Western Mail*, 12 April 1996).

> Thousands in all sectors of the beef industry are in real and far greater danger of catching diseases from which they will never recover. They are called ruin and bankruptcy and unless compensation materialises pretty fast the tragic consequences will leave lasting scars across the country's rural areas. Those who seek to preserve the structure of rural life should be aware that much more of the already depleted population of village people who work on the land will be lost forever. (Richards, 1996).

Secondly, the discourse of the farmer as the steward of the countryside was also invoked, this time to suggest that the financial consequences of BSE threatened not just the rural economy, but also the rural environment. Typical of this were comments by the deputy chair of the Farmers' Union of Wales in an article for the *Western Mail* newspaper:

> The beauty of the Welsh countryside is the result of sustained stewardship by the farming community and any serious decline in farm numbers must, inevitably, threaten the unique quality of the mountains and valleys of Wales ... The spectre now hanging over rural Wales in any serious decline in farming's fortunes is a terminal haemorrhage of our rural areas which could change forever the character of the Welsh countryside. (Thomas, 1996).

Thirdly, it was suggested that the publicity afforded to BSE and the cull of cattle could change public perceptions of the countryside. In particular, the association of rurality and purity (as contrasted to the dirt and grime of the city) was thought to be under threat as the media reproduced images of the countryside as a diseased space. Thus, it was feared, tourism could affected as visitors stayed away from rural areas, concerned about the risk to human health:

> We must do everything we can to support farmers and help them win back their customers because there is an enormous amount at stake – not least the image of Somerset as a safe and healthy place for people to come on holiday. (Leader, Somerset County Council, quoted in the *Somerset County Gazette*, 29 March 1996).

As the last quote indicates, the strategy had some success in shaping political responses at a local scale within rural areas, but its impact on central government policy was more limited. The eradication strategy based on the slaughter of potentially infected livestock continued to be pursued, and farmers failed to win the level of compensation that they had been pressing for. One consequence was to leave affected farmers more vulnerable to the series of further financial pressures that hit British agriculture in the following years. From 1996 the sense of crisis that had been provoked in beef farming by the BSE scare was replicated across different agricultural sectors, amounting to general agricultural recession. By August 1999 prices for agricultural produce had slumped so far that *The Independent* newspaper declared that farming was facing its 'worst crisis since the Thirties':

> farmers are losing cash on virtually everything they grow, breed, fatten and pick, from cabbages, potatoes, apples and oilseed rape to pigs, sheep, cows, eggs, chicken and milk. (*The Independent*, 28 August 1999, p. 1).

The unique coincidence of falling prices across a diverse range of agricultural produce (see table 6.1), was produced a combination of factors. Whilst the beef sector was still suffering from the loss of consumer confidence in the BSE scare and the continuing ban of the export of beef products, the strong value of the pound hit exports of other farm produce and intensified competition for domestic sales as imported food became comparatively cheaper. Moreover, as supermarkets competitively cut prices to attract customers the effect was to squeeze the 'farm-gate' price paid to farmers (table 6.2). At the same time, economic recessions in

Russia and the Far East of Asia hit export markets for by-products such as sheepskin, thus increasing farmers' dependence on food produce, and, more particularly, on subsidy payments. Indeed, arable farming was the only significant sector of the industry to escape economic meltdown largely because of the scale of EU subsidies for grain production.

Table 6.1 Change in average farmgate price paid to farmers in Britain

	1996-98	1997-99
Pig meat	- 60%	- 32%
Eggs	- 36%	- 24%
Lamb	- 36%	- 31%
Beef	- 35%	- 5%
Feed wheat	- 31%	n/a
Chicken	- 23%	n/a
Milk	- 22%	- 19%

Source: The Guardian, MAFF.

Table 6.2 Average share of retail price received by farmers

	1997	1999
Lamb	44%	32%
Beef	31%	30%
Pigmeat	36%	28%
Eggs	27%	21%
Milk	36%	30%

Source: The Guardian, MAFF.

Between 1994/5 and 1999/2000, the average farm income fell by 90 per cent. In the single year from 1998/9 to 1999/2000, the average farm income fell by 28 per cent. These average figures disguise even more severe collapses in particular sectors and particular regions. For example, in 1999/2000, the average farm income across the UK as a whole was £9,500, yet in Wales, it was just £4,500. The Llandovery branch of the NFU reported in October 1998 that most of its members qualified for Family Credit payments, in other words that they had an weekly income of less than £79. In real terms, many farmers were losing money against the cost of production. Estimates quoted by *The Independent* in August 1999 suggested that pig farmers were losing £11 on every pig raised, and potato farmers were losing up to £27 on every ton of potatoes produced. Other estimates suggested that sheep farmers were losing up to £15 per lamb. With so little to be made from the products of farming, the sale value of livestock also collapsed. A symbolic nadir was reached in September 1998 when lambs were sold at Abergavenny market for just 25 pence each.

The response of the farming unions was to embark on a discursive strategy aimed at attracting political support for the beleaguered farming industry. As such the representations of agriculture and rurality mobilized during the BSE scare were reactivated. These articulations once again attempted to construct the problems faced within agriculture as a crisis for the countryside as a whole by suggesting that the rural landscape valued by non-farming residents and urban visitors was dependent on agriculture:

> People watch programmes like Peak Practice and see the surroundings up here. It is beautiful countryside, but as my father said: 'You can't live on the view'. The irony is that this land would not be as beautiful if it were not for the farmers. (Derbyshire farmer, quoted in *The Guardian*, 2 February 2000).

At a UK level, this discursive strategy achieved only limited success. It secured some media coverage, but public opinion remained sceptical. Moreover, the problems of agriculture became conflated with the message of a more general 'rural crisis' being pushed by pro-hunting campaigners not least in the response of the UK government, which sought to challenge and contest the claims being made by the farming unions rather than to act on them. However, the strategy was also able to engage with the emerging politics of devolution in Britain, where it was able to achieve more purchase, particularly in Wales. The geography of Wales means that the countryside has a greater visibility in the national consciousness than in England, and rurality has always been a core element in Welsh national identity. As Gruffudd (1995) records, a traditionalist ruralism was a fundamental part of early Welsh nationalism in the 1920s and 1930s, and even in the immediate post-war years, Plaid Cymru advocated a policy of deindustrialization for Wales and the resettlement of industrial workers in 'agrarian colonies' in the countryside. The concentration of Welsh speaking heartlands in rural areas has further reinforced the cultural significance of rural life in Wales, and in particular, the significance of the traditional model of the 'family farm'. Thus, as the bias towards upland, livestock, farming in Welsh agriculture meant that it was particularly exposed to the agricultural recession, these conditions enabled the problems of farming to be easily represented not only as a rural crisis, but also as a national crisis.

The manufacture of this discourse of a rural crisis coincided with a period of political maneuvering ahead of the first elections to the National Assembly for Wales in May 1999, in which parties and interest groups jockeyed to dictate the landscape of the new Welsh political culture. As aspects of agricultural policy had already been devolved to the old Welsh Office, the Farmers Union of Wales and NFU Cymru were already organized at a Welsh level, and used to lobbying at that level, giving them an advantage over other groups who might have sought to represent the interests of rural Wales differently. Moreover, the message of a rural crisis in Wales precipitated by falling farm incomes was eagerly seized upon by the Welsh media, notably the *Western Mail* newspaper, keen to identify distinctively Welsh political stories, and by politicians of all parties who wanted to disprove the suspicion that the Assembly would be dominated by Cardiff and the South Wales valleys. Thus, by the time of the Assembly elections in May 1999, rural Wales had already been constructed as a 'space in crisis' (see Woods, 2005b). Agriculture was the single most debated item within the Assembly during its first year, and the discursive connection of the agricultural recession with a 'rural crisis' and a 'national crisis' was reproduced by Assembly Members:

> The increasing number of farms being forced out of business will in turn have implications for the whole social structure of rural Wales, the strength of the Welsh language in its traditional heartlands and the environmental sustainability of the countryside. (Chair of the Welsh Assembly Agriculture Committee in a letter to the Prime Minister, March 2000).

The effect of this was that unlike its counterpart in London, the Labour government elected in Cardiff – largely on the basis of urban votes – was not able to successfully challenge the idea of an agriculturally-based rural crisis or to assert its own representation of rural interests. Indeed, its one notable attempt to do so was spectacularly counter-productive. In selecting his first cabinet, the First Secretary, Alun Michael, appointed as Agriculture and Rural Development Secretary the AM for Carmarthen West and Pembroke, Christine Gwyther. As Alun Michael has himself argued subsequently, Christine Gwyther was uniquely qualified for the position as the only Labour AM with professional experience of rural development. However, as a non-farmer and a vegetarian, she immediately became a target for farmers' protests and a scapegoat for the problems in agriculture. Fewer than ten days after her appointment, the president of the Farmers' Union of Wales (FUW) had forecast that pressure from the industry would eventually force Christine Gwyther's resignation:

> It may not be tomorrow but the demand will gather pace. The general consensus is that appointing a vegetarian minister of meat has made Wales a laughing stock in the eyes of the world. Farmers are struggling and the outlook is none too bright. To think that the FUW from the start supported the Assembly and we want it to be strong and successful. We want an Agriculture Secretary who can lead from the front and tell the world how good all our produce is, whether it is beef, lamb or vegetables. (President of the Farmers' Union of Wales, quoted in the *Western Mail*, 24 May 1999, p. 1).

Gwyther withstood persistent hounding by the farm unions for fourteen months before eventually being dropped by the new First Minister, Rhodri Morgan, on the eve of his visit to the Royal Welsh Show, Wales's largest agricultural show. To this extent, the strategy of the farm unions may be judged to have been relatively politically successful in Wales, in succeeding in hijacking the representation of rural Wales and defining the job of the Agriculture Secretary in their own terms, positioning agriculture as a key political issue in Wales, achieving substantial media coverage, and eventually forcing a ministerial resignation. However, these achievements were essentially symbolic and did little to actually improve the situation of farming in Wales. It was hence in growing frustration at the apparent impotence of the established farm unions that farmers in Wales and elsewhere began to turn to more militant forms of direct action.

The New Agrarian Militancy

The history of farmer militancy in Britain is sporadic and generally limited. Unlike countries such as France or even the United States, where protests by farmers have been part of the landscape of rural politics, occurrences of farmer militancy in Britain have been the exception not the rule – demonstrations against the Labour government in the 1930s and protests by Welsh farmers in opposition to the introduction of milk quotas in 1984 being the most notable examples. Instead, the interests of farmers have conventionally been well represented by the farm unions operating through the agricultural policy community, thus nullifying the need for protests. At the same time, the culture of deference within the farming community (Newby, 1977) and the discourse of agrarian Conservatism (see chapter four), led farmers to construct themselves as peaceful, law-abiding, responsible businesspeople, whose approach to politics contrasted with the confrontational tactics of industrial trades unions. It therefore required a combination of the growing desperation of many farmers at their own financial situation, and a collapse in confidence in the ability of the farming unions to deliver an adequate response, for a new strain of agrarian militancy to be sparked in the late 1990s.

The tipping point came in November 1997 in north Wales, among farmers who were frustrated that whilst a ban on the export of British beef was still in place, and whilst there appeared to be little action on the part of the British government to respond to the growing wider recession in agriculture, imports of food stuffs were not only permitted, but often had a commercial advantage due to the strength of the pound. On 30 November 1997, around 400 farmers meeting to discuss their problems made a spontaneous decision to go to Holyhead docks to 'reason' with the drivers of lorries carrying imports of Irish beef. During the course of the night one small group of farmers broke into the trailer of an Irish lorry to discover a consignment of Irish beefburgers for delivery to Tesco supermarkets, and tipped forty tonnes of the burgers into the sea.

The Holyhead demonstrations generated significant media coverage and set a precedent for further direct action. Not only did protesters return to Holyhead on subsequent evenings, but copycat protests were initiated in other parts of the country. By the end of the first week, port blockades had also been mounted at Fishguard, Swansea, Portsmouth, Heysham, Stranraer, Plymouth and Dover, and at the Channel Tunnel in Folkestone; and pickets mounted at a range of other strategic sites including supermarkets and distribution depots (see figure 6.1). Farmers and their supporters had also mounted demonstrations in the town centres of Abergavenny, Newtown and Monmouth. Significantly, the protests were condemned by the farm unions, but with little effect. Both the FUW and NFU belatedly responded to grassroots pressure by organizing rallies in London, attended by 2,000 farmers and 3,500 farmers respectively. But similarly large numbers of farmers also continued to participate in the direct action protests. As

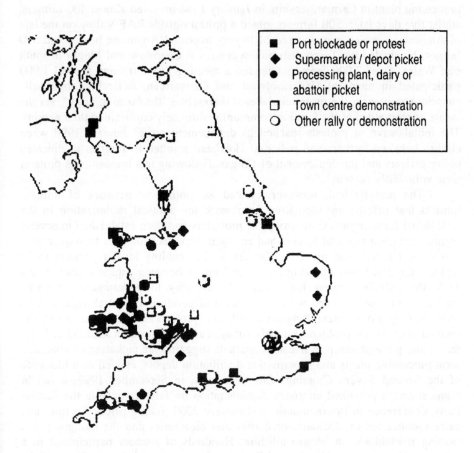

Figure 6.1 Farmers' protests in Britain, December 1997-March 2000

Source: Farmers' Weekly, Western Mail, The Guardian.

well as the 400 protesters at Holyhead, around 200 farmers joined protests in Fishguard and around 300 in both Swansea and Plymouth. A blockade of a meat processing plant in Carmarthenshire in January 1998 involved almost 400 farmers, whilst five days later, 500 farmers joined a protest outside RAF Valley on the Isle of Anglesey – which was accused of buying imported Argentine beef – and 500 farmers blockaded the Tesco distribution centres at Chepstow and Magor in south east Wales. Over 2,000 farmers attended a meeting Oswestry and around 1,000 participated in meets at Haverfordwest and Carmarthen. Reflecting their self-conscious militancy, the protesters labeled themselves 'the farming army', yet the innate conservatism of the farming community ultimately constrained their actions. The initial wave of protests reached its denouement on 27 January 1998 when clashes between farmers and police at Holyhead resulted in injuries to nineteen police officers and the deployment of CS gas. Following this incident, the protests were voluntarily ceased.

The protests had, however, created an embryonic network of militant farmers that offered an alternative framework for political mobilization to the established farm unions. Core groups of protesters had been established in several regions and clear regional leaders had emerged. New technologies such as mobile phones and e-mail had proved to be key tools, enabling local protesters to be rapidly mobilized and allowing for co-ordination between regional leaders. As such, the militant farmers had gained the capacity to undertake a 'guerrilla campaign' of protests, even without any kind of permanent, central organization. Over the course of the next thirty-six months, the protests and demonstrations were revived as economic problems hit different agricultural sectors. In September 1998, the falling price of sheep in livestock markets triggered the picketing of abattoirs, meat processing plants and supermarket distribution depots, as well as a blockade of the Second Severn Crossing. One year later, in September 1999, a fall in pigmeat prices provoked an angry demonstration by farmers outside the Labour Party Conference in Bournemouth. In February 2000, falling prices for milk and dairy produce led to blockades of dairies and creameries and the mounting of a 'rolling road block' in Monmouthshire. Hundreds of tractors participated in a demonstration in central Cardiff, whilst other farmers joined a 'Fair Share of the Bottle' march from Carlisle to London. By March 2000, *The Farmers Weekly* had calculated that a total of 102 protests had been mounted by farmers since 1996 (Reed, 2004).

The pattern of participation in such protests, however, was complex. As Reed (2004) shows with reference to the picketing of dairy plants in the south west of England, apart from a core group of protesters many demonstrators were one-time participants whose involvement was constrained by the time demands of farming and by a hesitancy at the act of protesting. In many cases, the decision to protest was the result of personal desperation and an individual rationalization that the demonstrations would generate pressure that would help the NFU to secure deals that would benefit farmers. Thus, whilst the NFU continued to distance itself as an organization from the protests, senior local NFU members were among the occasional participants.

The loose network of protest cells was formalized with the creation of an over-arching pressure group, Farmers for Action, at a meeting of regional protest leaders held at a motorway service station in May 2000. The establishment of Farmers for Action (FFA) gave the militant farmers a more formal identity, a national figurehead in the form of FFA chair, David Hanley, a formal membership and a website, but it did not substantially alter their tactics. Farmers for Action did not seek to engage in political lobbying alongside the NFU and FUW, but rather came into existence as a co-ordinating body for local protest groups, drawing inspiration from militant French farmers (particularly the *Confédération Paysanne*) and radical environmentalist groups. As such, Farmers for Action may be perceived to be engaged in a 'postmodern politics of resistance' by mounting symbolic challenges and rendering power visible (Routledge, 1997; see also chapter five).

Farmers for Action does not seek to be incorporated into the processes of government. Some of its protests are intended to be directly disruptive or obstructive, but many are symbolic actions. The targeting of supermarkets, food companies and other corporations is clearly an attempt to render visible and negotiable the power relations impacting on agriculture – far more so than the continuing focus of the established farm union on government departments and politicians. Moreover, the need to attract public attention is an important part of rationality of many of the demonstrations, including media-friendly stunts such as dumping 50,000 litres of waste milk on to fields with muckspreaders, and setting a group of pigs free in Parliament Square.

However, in order to maintain media interest, protests need to be novel and newsworthy. Here again the political naivety and hesitancy of the farmers became an obstacle. In June 2000, a BBC Wales documentary followed two Farmers for Action leaders, Brynle Williams and David Hanley, preparing for a demonstration in central Manchester against the high price of fuel which was compounding the financial situation of many farmers. The demonstration was planned to minimize the disruption caused to shoppers, and with the full co-operation of the police. Consequently, most shoppers in Manchester city centre on the day failed even to notice the demonstration and it attracted no media attention. Towards the end of the documentary, the FFA leaders are shown discussing their frustration at the lack of publicity. In a revealing remark, Brynle Williams reluctantly suggests that it maybe time to 'copy the French' and step up their activities to bigger demonstrations that would directly inconvenience the general public.

The Fuel Protests

The campaign against high fuel prices, and more specifically, the high level of tax on fuel, had been developed over the summer of 2000 by a loose coalition of actors including not only Farmers for Action, but also hauliers and right-wing newspapers. A 'Dump the Pump' consumer boycott campaign had been promoted, but with little impact. Fuel prices were also an issue elsewhere in Europe and on 18 August, French fishermen, hauliers and farmers started a blockade of ports and oil depots, eventually winning tax concessions from the French government (Mitchell and Dolun, 2001). As the French blockades came to an end at the beginning of September, similar protests were launched in a number of other countries including Belgium, Germany, Ireland, Italy, Norway, the Netherlands, Spain and Sweden. In each case, the protesters were successful in winning concessions on fuel tax (Mitchell and Dolun, 2001). This achievement was not lost on British farmers, as one FFA spokesperson acknowledged,

> We looked at the French and we thought enough is enough. We just cannot survive. We have got to get the price of fuel down. (Paul Ashley, quoted in *The Guardian*, 8 September 2000).

On 7 September 2000, a hurriedly convened meeting of 160 farmers at St Asaph in North Wales resolved to organize an immediate blockade of the Shell oil depot at Stanlow in Cheshire. As Doherty *et al.* observe,

> The decision to mount this protest was essentially spontaneous. By this we do not mean that it was an irrational or instinctual action, but rather an on the spot decision, the consequences of which were unanticipated. Current social movement theory tends to over-emphasise the strategic aspects of action at the expense of spontaneity. Some who went to the meeting in St Asaph were prepared to take some kind of protest action, but the decisions to act immediately and to target the fuel depot were unplanned and unanticipated. (Doherty *et al.*, 2003, p. 8).

In the same manner as the earlier farmers' blockades, the action at Stanlow was quickly copied around the country. Within forty-eight hours blockades had also been mounted at Avonmouth, Pembroke and Milford Haven, spreading the following day to oil depots in Manchester, Hampshire and Lincolnshire. As Robinson (2003) notes, 'by Monday [11 September] most of the country's refineries and depots were effectively closed' (p. 424). The staunching of the supply of fuel was compounded by panic buying, such that 90 per cent of petrol stations were estimated to have run out of fuel by 13 September (Robinson, 2003). With knock-on consequences for food distribution, commuting and the operation of the emergency services, the situation rapidly escalated into a national political crisis, heavily stirred by the media. From an initial position of denial of the scale of the problem, the government's strategy moved swiftly to an assertion of 'business as usual', to crisis management to a final phase in which the fuel protesters were portrayed as 'enemies of the state' as the government began to reclaim control of the situation (Robinson, 2003).

The role of Welsh farmers in initiating the protests, and the prominence of Brynle Williams and David Hanley as spokespersons for the protesters, served to identify the blockades with the farming community. Moreover, the public was receptive to the apparent logic of the argument that fuel tax had differential impact in rural areas where car use was a necessity for many. Mitchell and Dolun (2001) note that 'countryside commuters' joined blockades in Milford Haven, Bristol and Manchester. However, the fuel blockades exhibited a number of differences to the earlier farmers' protests. They involved, for instance, much smaller numbers of protesters than the earlier demonstrations. Only around 150 protesters participated in the largest blockades at Stanlow, Milford Haven, Pembroke and Kingsbury in Warwickshire, and generally only 60 to 70 protesters were active at other sites. Furthermore, many of these protesters were not farmers. Mitchell and Dolun (2001) suggest that hauliers were the single largest group involved, particularly at sites in the east of England, and that taxi drivers and fishermen also joined some blockades. Finally, the role of the oil companies themselves remains an enigmatic issue. With relatively few protesters participating in most of the blockades, suspicions developed that the oil companies were deliberately holding back fuel deliveries for their own ends. As one senior government advisor commented,

> There's something not quite right about it. It's not a normal demonstration and they should get their sleeves rolled up and start delivering the oil where it belongs, at the petrol stations. There does seem to be some collusion with the protesters. Perhaps it's because if the fuel duty were reduced they would sell more petrol. (Lord Mackenzie, quoted in *The Guardian*, 13 September 2000).

Thus, although farmers were in many ways the lynchpin of the fuel protests they lost control of the protests as they progressed, with many becoming increasingly uneasy at the scale of the political crisis that had been generated. Differences emerged between farmers and the more militant hauliers, as well as between the leaders of FFA, notably Brynle Williams and David Handley, about how far the protesters would be prepared to go. Underlying these tensions remained both the political inexperience of the demonstrators, and their inherent conservatism. Williams had, according to his own description of events, set out a moderate course at the beginning of the protests, telling demonstrators before the blockade of Stanlow:

> You have to remember what you are getting yourself into: you are not challenging a commercial company. You are challenging the Government. Be under no apprehension. This is a very, very serious thing that you do ... Emergency services get total priority. You don't hinder them in any shape or form. And the first senseless act of violence and that's it. It's over. You are all businessmen in your own right and hopefully gentlemen too, and will act accordingly. (Brynle Williams, quoted in Treneman, 2000, p. 4).

As the effects of the protests grew, this wariness mushroomed into unease not that the protests may fail, but that they may achieve too much. It was thus as reluctant revolutionaries that the protest leaders decided to prematurely end the blockades in favour of setting a sixty day ultimatum to the government to cut fuel tax. As Williams again explains,

> It was in the interest of common sense and decency. We weren't there to topple a government. We were there to demonstrate our problems. And we walked away. We walked away as businessmen. Not as bloody hooligans. Not as vandals. Not as anarchists. Not as leftists, rightists, centrists, whatever. (Brynle Williams, quoted in Treneman, 2000, p. 4).

Significantly, when the hauliers, under the banner of the Peoples' Fuel Lobby, attempted to resurrect the protests after the sixty day deadline had expired – and again on later occasions – they failed to attract public support and to achieve a political impact. The tensions that emerged during the fuel protests also served to further polarize opinion within the farming communities, as symbolized by the divergent paths of the most prominent leaders, Brynle Williams and David Handley. Whilst Brynle Williams moved into electoral politics, becoming a Conservative Member of the Welsh Assembly in the 2003 elections, David Handley has continued to lead Farmers for Action, which has consolidated its position as a radical voice for farmers, operating through a diverse range of direct action protests.

Foot and Mouth

On the 1 January 2001, *The Western Daily Press* carried a front page editorial calling for 2001 to be 'the year of the countryside'. With a message that would appeal to rural readers in its west country circulation area, the newspaper argued for politicians to address the problems of the countryside in the wake of the fuel protests, continuing economic difficulties in agriculture, and continuing uncertainty over the future of hunting. It turned out to be an ironic prophecy. For eight months of 2001, the countryside did indeed dominate domestic news coverage and political debate in Britain, but not in a way envisaged by the *Western Daily Press*. On 19 February, a veterinary inspector at a abattoir in Brentwood, Essex, identified suspected cases of Foot and Mouth disease (FMD) in pigs destined for slaughter. Within hours the incidence of FMD had been confirmed and restrictions were imposed around the farms from which the pigs had been sourced. It was already too late, however, to stop the spread of the epidemic, which was later confirmed to have originated at a pig farm in Northumberland. By 23 February, six outbreaks had been confirmed. By 26 February – one week after the initial identification – eleven suspected outbreaks had been reported or confirmed in locations as far apart as Essex, Wiltshire, Herefordshire and Devon. The geographical spread of these outbreaks should have provided a warning about the difficulty that would be faced in containing the epidemic. Modern farming practices, including the transfer of livestock through different markets across the country, the long-distance transportation of livestock to abattoirs and high stock densities on farms, helped to create conditions in which the virus could spread rapidly and extensively. Confirmed cases of FMD increased each day from the end of February to a peak of fifty cases per day in late March. In total, Foot and Mouth Disease was confirmed at 2,030 premises, making it the world's worst ever recorded epidemic of FMD.

A worldwide ban of the export of British livestock, meat and animal products was imposed by the European Commission on 21 February, and remained in force until Britain was officially declared FMD-free in January 2002. In order to eradicate the disease, the Ministry of Agriculture, Fisheries and Food (MAFF) employed a strategy based on that used during the last major FMD outbreak in Britain in 1967. Quarantine zones were imposed around all premises with suspected cases and livestock on these premises and contiguous farms were slaughtered. In total, just over four million animals were culled at 9,503 farms. A further 1.7 million animals were culled under a welfare scheme due to problems created by the prohibition of livestock movement, whilst a further two million lambs, calves and piglets are estimated to have been slaughtered shortly after birth but not included in official figures. As such, the total number of animals killed as a result of the outbreak has been estimated to be around 7.7 million, or one eighth of all farm animals in Britain (Private Eye, 2001). Other measures introduced to try to control the spread of the epidemic included road closures near infected premises, the blanket closure of public footpaths in most rural counties, and the cancellation of many events, including a march planned by the Countryside Alliance for March 2001. The total cost of the exercise to the taxpayer has been estimated at £2 billion, including £1.75 billion paid in compensation to farmers whose livestock had been slaughtered.

The scale of this impact is extraordinary given that, like BSE, the Foot and Mouth epidemic essentially presented a problem of disease control that directly affected only the relatively small livestock farming industry. Unlike BSE, Foot and Mouth does not in itself pose any threat to human health. It can be fatal in cattle and sheep, but its main problem is that it severely reduces the productivity of infected livestock. Most significantly, it is a highly contagious virus, not least because it can jump species and can infect any cloven-hoofed animal. As such, FMD-free countries are extremely protective of their status and tight trade restrictions are imposed on any country where Foot and Mouth is suspected. Indeed, the protection of the British agricultural export industry appears to have been the top priority in the response to the epidemic, with the key argument used against the use of vaccination being that it would prolong the period before Britain could be declared as FMD free and therefore prolong the export ban.

However, the Foot and Mouth epidemic did not remain a problem of disease control, but became successively a political crisis; an economic crisis effecting not only agriculture but also the wider rural economy, particularly tourism; and a local social crisis in the worst affected areas. In so doing, the epidemic revealed both the continuing influence of the NFU within the Ministry of Agriculture (MAFF), and the extent to which the model of the rural economy employed by these two actors had become disconnected from the realities of the contemporary countryside. Crucial to the mismanagement of the outbreak was the way in which the NFU/MAFF nexus apparently still conceived of the countryside as a primarily agricultural space. As such in constructing FMD as an agricultural crisis, they implicitly also constructed it as a rural crisis, and established a rationale that the protection of agricultural interests would outweigh any potentially negative consequences of the control strategy. Yet, as Ward *et al.* (2004) observe, 'framing FMD as purely an agricultural problem meant that the control strategy was just that demanded by the agricultural community. Other interests were marginalized' (p. 297). From the beginning of the crisis the NFU was afforded what one minister described as 'almost open door access' (Elliot Morley, quoted in Ward *et al.*, 2004, p. 297), enabling it to exert considerable influence over the strategy adopted by the Ministry. This close relationship built on the historical precedent of the agricultural policy community and excluded other groups in the same way that the policy community had done – by defining them out of the picture. In the rationality of the NFU/MAFF nexus, along as Foot and Mouth could be conceived of as an agricultural crisis there was no need to consult other groups, including rural tourist operators, ramblers or even small farmer organizations.

Moreover, the initial handling of the Foot and Mouth epidemic again exposed the uncertainty of the New Labour government in dealing with rural, and particularly agricultural, issues. Labour ministers and advisers lacked the confidence and the knowledge to challenge the NFU/MAFF representation of the problem and thus became dependent on the 'expertise' of MAFF civil servants, the NFU, and their scientific advisers, even though several of the assumptions mobilized were later shown to be questionable (Anderson, 2002; Bickerstaff and Simmons, 2004; Ward et al., 2004). Fundamentally, the plan of action developed by MAFF in consultation with the NFU contained three major errors. Firstly, it miscalculated the extent to which agriculture had changed since 1967 and in particular the scale of animal movements around the country. This meant that the control measures being implemented were failing to keep up with the spread of the virus. Secondly, it failed to reflect the structure of post-devolution governance, in which agricultural policy had been devolved to the administrations in Cardiff, Edinburgh and Belfast. Whilst the Scottish and Northern Ireland governments were generally considered to have responded well to the outbreak, in Wales action was compromised by the fact that the Assembly had been devolved the responsibility but not the appropriate resources. In particular, the Welsh Assembly Government had no authority over the State Veterinary Service, which was responsible for implementing the disease control measures. Consequently, confusion was generated between London and Cardiff over the management of the crisis and *ad hoc* conventions had to be adopted as events developed. Thirdly, the plan misrepresented the contemporary rural economy, failing to understand the impact of the control measures on other economic sectors such as tourism and over-emphasizing the importance of agriculture to the rural economy. In summary, the Foot and Mouth epidemic was mishandled politically because individual elements, such as the virus or the rural economy, were incorrectly modeled, and the network of actors assembled to respond to the problem did not include all the appropriate actors. As such, Donaldson et al. (2002) have argued that:

> It is not mismanagement of FMD that created the crisis (although institutional deficiencies undoubtedly played a role in some botched translations), but a fundamental misapprehension of the actors involved. (Donaldson *et al.*, 2002, p. 211).

The Foot and Mouth epidemic was translated into a political crisis by the failure of the initial strategy adopted by MAFF to control and restrict the impact of the outbreak. As a consequence of the strategy, confrontation was provoked not only between the government and farmers, but also between the government and the tourist industry, ramblers, animal welfare campaigners and rural residents. Furthermore, the apparent inability to control the spread of the epidemic led to a questioning of the government's competence. With the government planning to call a general election for 1 May (subsequently delayed until 10 June), these unintended consequences posed a major political risk. It was hence the political threat that led Tony Blair to redefine FMD as a national crisis, engaging on a crisis management strategy that removed command of the control strategy from MAFF and located it within the Cabinet Office Briefing Room (COBR). This action, taken during the week of 21-27 March, signaled that FMD was not just an agricultural problem.

From the beginning of April onwards, the protection and promotion of tourism became a key priority in the management of the crisis alongside the existing agricultural priorities. The economic crisis in tourism, however, was a self-inflicted problem. FMD was always going to produce an economic crisis in agriculture and the 'Lessons to be Learned' Inquiry estimated that the cost to agricultural producers was around £355 million, of which £130 million resulted from the loss of exports (Anderson, 2002). Foot and Mouth only became a problem for the tourist industry when local authorities began to close public footpaths – under pressure from the government and farming unions – and the message was broadcast that people should stay out of the countryside. The immense pressure that was placed on the public to voluntarily restrict their movement in rural areas and used to justified hurried legislation to permit blanket footpath closures was based at best on unqualified assumptions. Virtually all the spread of FMD could be accounted for by animal movements (and later by the movement of contractors engaged in tackling the epidemic), and the 'Lessons to be Learned' Inquiry noted that action was taken 'not as far as we have been able to determine on the basis of explicit veterinary and scientific advice' (Anderson, 2002, p. 64). As the Inquiry report goes on to conclude:

> The frequent changes in guidance, the lack of clarity in communication, the loss of confidence in the Government's scientific understanding and control of the outbreak, all hindered those seeking to get consensus on reopening. Farmers under tight biosecurity restrictions found it difficult to accept that walkers would not pose a risk. The public began to wonder why a disease that reportedly did not have much of an impact on animals and for which a vaccination was said to exist should be causing so much disruption. Only with the establishment of the Rural Task Force in mid March did a strategy begin to emerge for reopening footpaths based on a process of veterinary risk assessment. There was greater appreciation of the need to get the balance right between rural access and disease control. But, in terms of the wider economic impact on tourism and the rural economy, the damage had been done. (Anderson, 2002, p. 64).

The rural tourism industry, worth £12 billion per year, was reported to be losing around £100 million each week by the end of March (*The Guardian*, 31 March 2001). In areas such as the Lake District, tourist visits over the Easter weekend were reported to be down 80 per cent on normal (*The Guardian*, 16 April 2001), but it was not just rural tourism that was affected. Urban tourist attractions were also hit as overseas visitors cancelled trips to Britain. In total the direct cost to the tourism sector was estimated to be between £2.7 billion and £3.2 billion

(Anderson, 2002). Overall, the Foot and Mouth outbreak and its consequences were estimated to have cost between £2.2 billion and £2.5 billion to the rural economy. Interestingly, this impact had been more or less accurately predicted by the Countryside Agency at the beginning of March, but the agency had experienced opposition from MAFF when they published the projection in a press release (Anderson, 2002).

Aggregate national figures do not, however, convey a full picture of the geographically differentiated impact of FMD. Of the 2,030 confirmed outbreaks of FMD, two-thirds (1,375 cases) were concentrated in the four counties of Cumbria, Dumfries and Galloway, Devon and North Yorkshire (see Table 6.3). A further 523 cases were concentrated in ten more counties, mostly in the north of England or along the English/Welsh border. Within these worst-affected regions, particular local crises developed as a result of the cumulative impact of disruption to both economic and social life. Economically, the impact in the worst-affected regions was greater not only because of the number of farms involved and the virtual collapse of local tourism, but also because the suspension of normal everyday activity. In Powys, for example, the local economy was estimated to have experienced a 64 per cent drop in annual profits as a result of FMD. Around 58 per cent of the lost profits were in the retail, distribution and catering sector, with hotels, guesthouses and self-catering accommodation all essentially returning a trading loss for 2001; but significant falls in profits were also reported by businesses in agriculture, manufacturing, transport and communications and other services (Rural Wales, 2002; see also Scott *et al.*, 2004). A total of 470 businesses in the county closed during 2001 because they found it impossible to continue operating against the impact of FMD.

Table 6.3 Premises infected with Foot and Mouth, by county

	Number of premises	As % of all cases
Cumbria	893	44.0
Dumfries and Galloway	176	8.7
Devon	173	8.5
North Yorkshire	133	6.6
Northumberland	88	4.3
Durham	85	4.2
Gloucestershire	72	3.5
Powys	70	3.4
Lancashire	53	2.6
Herefordshire	49	2.4
Staffordshire	45	2.2
Monmouthshire	23	1.1
Worcestershire	22	1.1
Cheshire	16	0.8
Others	132	6.5

Source: DEFRA.

Alongside the economic crisis, a social crisis was generated in many localities by the measures imposed to try and restrict the spread of the disease. Problems of isolation were greatest, of course, for households on infected farms who were quarantined on their own premises (and other farm households who imposed a voluntary quarantine in an attempt to protect themselves from the epidemic), as detailed by a number of diary accounts (see for example Leaney, 2001). Yet, social and psychological problems were also produced at a collective level by the suspension of most community activities and the virtual curfew conditions that were self-imposed in many rural communities. In several areas, local support groups were set up in order to try to respond to problems of stress, depression and business failure.

The local support groups were one of several types of grassroots political action that emerged from the Foot and Mouth crisis. In some cases political mobilization reflected anger towards the government and articulated a feeling of abandonment. Government ministers were picketed by groups such as the Cumbria Crisis Alliance on visits to infected areas and slogans such as 'Tony Blair – he don't care' appeared on farm building roofs, hay bales and roadside signs. A second focus of political mobilization was the cull of livestock who had not been infected by the virus but who fell within the infected area. Public protests were mounted at a number of sites such as Oaklands Park in Gloucestershire, a Steiner community and organic farm where MAFF's intention to slaughter the healthy livestock was successfully resisted. Finally, political mobilization also occurred in response to proposals for the disposal of slaughtered livestock. In Wales, the identification of the Epynt Army Ranges west of Brecon, as the site first for an incineration pyre for carcasses and later for burial pits for slaughtered livestock, provoked protests by local residents. On 1 April, demonstrators blocked the A40 road at Sennybridge to stop lorries carrying straw and wood for the pyre to the site. The following day, police officers were injured when a bulldozer was driven into a police van during a protest involving 300 participants (Cook, 2001). Although unsuccessful in preventing the pyre, protesters continued to oppose plans for the creation of burial pits, displaying protest banners (see figure 6.2) and seeking a legal injunction. The plan for the burial pit was later dropped, officially because there was insufficient need.

The Foot and Mouth epidemic proved to be a watershed event for agriculture in Britain. At one level, within infected areas, sympathy for farmers who saw their stock destroyed helped to rebuild connections between farming and the rural community that had been eroded by social and economic restructuring. Yet, at a broader level, the Foot and Mouth 'crisis' marked the end of farmer exceptionalism in government. Following his re-election on 10 June, Tony Blair moved swiftly to abolish the Ministry of Agriculture, Fisheries and Food (MAFF) and subsume its responsibilities within a new Department for the Environment, Food and Rural Affairs (DEFRA). The absence of any explicit reference to agriculture in the title of the new department was highly symbolic, signifying that the countryside was to be constructed in new terms. As the DEFRA Minister Michael Meacher, argued in a revealing letter to *The Guardian*, the creation of the new department was viewed as 'an inspired seizure of the best opportunity for

158   *Contesting Rurality*

reform in decades' (Meacher, 2001, p.15). The mood for change was shared by the public which had tolerated the restrictions imposed to deal with Foot and Mouth, but which anticipated that the farming community would make a gesture in return by changing unpopular farming practices. At the height of the crisis there was a widespread belief that the epidemic and the cull had created an unforeseen opportunity to restock with lower densities and to move towards a less intensive form of agriculture. Yet, the opportunity passed largely untaken. The availability of compensation payments encouraged farmers to restock exactly as they had done previously. Scott *et al.* (2004), for example, found that just under half of farmers surveyed in Wales intended to fully restock, and only around a quarter intended to restock at a lower density.

Figure 6.2 Protest against disposal of infected carcasses at Sennybridge, Powys

Conclusions: Farming Futures?

The future of farming in Britain into the twenty-first century now more than ever appears to be uncharted territory. As the old certainties of agricultural policy and practice have been undermined by the events of the last two decades, they have failed to be replaced by any firm indication of a coherent new policy direction. In part this reflects the incoherence of the contemporary agricultural policy network in Britain, with influence over policy-making dispersed between a number of different actors operating at different scales. The national agricultural policy regime, which had been established in Britain during the immediate post-Second World War years, effectively had its power curtailed with British accession to the European Economic Community in 1972. Since then, the Common Agricultural Policy (CAP) of the European Community/Union has been the primary power centre for British agriculture. Overall, this has been to the benefit of British farming as agriculture has maintained a privileged position with EU policy-making for far longer than it has within domestic politics. Thus, although episodes such as the BSE scares in Britain and Germany and the Foot and Mouth epidemic in Britain have helped to slowly build a coalition of member states that are favourable to CAP reform, substantial reform is still obstructed by countries such as France where agricultural unions continue to be more influential in domestic politics. Indeed, it is notable that the moderate reforms of the CAP introduced under Agenda 2000, including the ending of production-based subsidies, were provoked by fiscal concerns about the implications of EU enlargement into central and eastern Europe, rather than a reassessment of the state of agriculture.

The EU itself has also ceded some power over agricultural policy upwards to the World Trade Organization (WTO). The global regulation of agricultural trade has been a key issue in the Doha round of WTO talks, with pressure placed on the EU and the United States to cut back state support for agriculture by both the free-trade Cairns Group of leading agricultural exporters and fair trade advocates. The probable outcome of the negotiations, which is likely to include the withdrawal of export subsidies, will impact on British farmers. Yet British farmers seem to be less engaged with the WTO process than their French or American counterparts. At the same time, agricultural policy within the UK has been devolved to administrations in Scotland, Wales and Northern Ireland. Although all operate within the parameters of the CAP, they do now have discretion to interpret aspects of agricultural policy and, for example, will be implementing the new system of subsidies differently. Yet, arguably one of the most significant changes has been the increasing power of large corporations, particularly supermarkets and food processing companies, in the agri-food system.

This new framework of power in agriculture has required the development of a new agricultural politics, challenging both new and established actors. The NFU has maintained its position as the most prominent agricultural interest group in part because it has adapted to the multi-scalar nature of agricultural governance, working at EU level through the Comité des Organizations Professionelles Agricoles (COPA) and organizing semi-autonomously in Scotland and Wales. Similarly, Welsh devolution has created new opportunities for the Farmers Union of Wales, enabling it to engage with government on an equal footing with the NFU, something which it had never achieved in Whitehall. In contrast, smaller groups such as Farmers for Action do not have capacity to organize at these different levels, although government institutions and meetings at all scales may be targeted by their protests. Conversely, it is the newer groups such as Farmers for Action who have responded most explicitly to the power exercised by supermarkets and food companies. The type of action practiced by Farmers for Action, including blockades of supermarket depots and pickets at food processing plants, have a level of visibility and directness that is absent from the more discrete lobbying favoured by the established farming unions, and thus can greater impression of 'something being done'. This in turn has influenced farmers' perceptions of the different organizations. One survey of 533 small farmers in England, undertaken in September 2002 for a group critical of the NFU, found that 18 per cent had let their NFU membership lapse, and 28 per cent felt their interests were not being represented (Independent Farmers' Group, 2002). Criticisms of the NFU have been regularly expressed by leaders of Farmers for Action, often in strident terms, whilst the NFU in turn has been dismissive of FFA. As Grant (2004) observes,

> The NFU seemed to follow a strategy of ignoring the FFA, presumably in the hope that it would eventually fragment and disappear. However, the FFA has been particularly successful in drawing support from smaller scale dairy farmers, hard hit by reductions in the price for milk. It has continued its policy of blockading dairy processing plants and the distribution depots of retailers and has won some concessions from retailers on milk prices, although it encountered more resistance from the processors. (Grant, 2004, p. 413).

Yet, as Grant continues to note:

> If the FFA continues to flourish, it would mean that farmer representation in Britain has been divided on the French model of more conventional and more militant organizations. That would be an indication of a declining attractiveness of a insider strategy. (Grant, 2004, p. 413).

In November 2002, a further organization aimed at representing small farmers was launched with the name 'Farm'. Backed by the environmental campaigner Zac Goldsmith, Farm has adopted a distinctly progressive agenda and sought to develop links with consumer groups. Farm has not engaged in direct action protests in the same manner as Farmers for Action, but has organized demonstrations and pickets, as well as running public relations campaigns, lobbying and commissioning research. In this way it has adopted tactics that are more akin to those of mainstream environmental organizations, or of the Rural Coalition in the United States. Like Farmers for Action, however, Farm has made the supermarket chains a prime target, for example by picketing the AGM of Tesco in June 2004. Yet, Farm has also been dismissed by the NFU, whose retiring chair was apparently criticizing the group when he told the NFU conference that 'standing on the sidelines and shouting has proved to get no-one nowhere' (www.farm.org.uk).

Tensions between the established farm unions and newer more militant groups became public in August 2002 around a 24 hour production strike organized by Farmers for Action. FFA and the established unions clashed over both the level of participation – FFA claimed that 70,000 farmers took part, the unions suggested that the number was far fewer – and the value of the tactic. The NFU described the strike as 'an extremely risky strategy' and warned that 'such action could damage the relationship between individual farmers and their buyers, and lead to long term financial losses for the farmer' (*The Guardian*, 24 August 2002). Similarly, the FUW – which had been criticized by FFA for not instructing its members to join the action – warned that the strike could put farmers' businesses in jeopardy. Yet, despite the public antagonism, tentative connections have been developed. Local branch officers of both the NFU and FUW have participated as individuals in demonstrations organized by Farmers for Action, and, as Grant (2004) notes, certain individuals within the unions' hierarchies have expressed greater openness towards FFA than the official line. More explicitly, the Scottish National Farmers Union has broken ranks with its English counterpart by collaborating with FFA in organizing blockades of supermarket distribution depots. The English NFU itself has embarked on a subtle repositioning by relocating its offices from London to Warwickshire. As Grant (2004) again observes, the move can be interpreted as 'a strategic decision to move closer to its members and the market-place and to downplay close links with government' (p. 414).

Thus the picture of twenty-first century agricultural politics that is emerging is one in which no organization possesses the status of a classic insider group, but in which several farmers' organizations are jostling for position and influence as outsider groups with more or less militant agendas. Whilst the organizations themselves compete publicly with each other, there are signs that individual farmers do not perceive them to be mutually exclusive allegiances. Rather, farmers seem to be constructing individual pathways of activism that suit their own personal circumstances, with may involve both participation in direct action protests and activity within the established farm unions depending on their assessment of the value of the activities concerned.

Chapter 7

Developing the Countryside: Discourses and Dissent

Public Attitudes Towards the Countryside

The politicization of the British countryside at the end of the twentieth century resulted not only from the social and economic restructuring of rural regions, but also from a concurrent shift in public attitudes towards the countryside. During the mid part of the twentieth century the representation of the countryside as an agricultural space had widespread acceptance within both popular and policy discourses. From the 1960s onwards, however, the strength of this discourse weakened, to be replaced by a discourse of the rural as a space of nature. Informed by new scientific knowledge that has highlighted the damage caused by modern agricultural practices, pollution and urbanization to the rural environment (see Woods, 2005a), but also by stereotypical representations of rural life and of animals in the media, the reproduction of this discourse has helped to generate scepticism towards agriculture and opposition to large-scale developments in rural space. As many of the reference points that inform the discourse of the rural as a space of nature are taken from global media representations or from globally-disseminated scientific ideas, this discursive shift can be positioned as part of a 'globalization of values' in which 'universal' standards and cultural expectations are imposed on rural spaces, displacing locally-grounded discourses of rurality and nature (Woods, 2005a). This process, to many rural activists, has resulted in the distortion of public opinion through misinformation and misrepresentation, as a former chief executive of the Countryside Alliance has suggested:

> a generation brought up on *The Animals of Farthing Wood*, Walt Disney films and visits to theme parks is easy meat for single-issue pressure groups who exploit this lack of understanding of the realities of the countryside to their own ends (Hanbury-Tenison, 1997, p. 92).

The British Social Attitudes Survey traced changing public attitudes towards the countryside during the 1980s and 1990s, revealing a lessening of sympathy towards agriculture and a strengthening conservationist instinct. Although around three-quarters of respondents continued to believe that farmers looked after the countryside well (76 per cent in 1985, 72 per cent in 1989), the proportion believing that modern farming damaged the countryside increased from 64 per cent in 1985 to 72 per cent in 1989, with the proportion agreeing that 'looking after the countryside is too important to be left to farmers' increasing from 34 per cent to 46 per cent over the same period (Jowell *et al.*, 1990). The 1987 survey also found that a majority of respondents thought that the countryside had changed for the worse, and that just under half (44 per cent) said that they were 'very concerned' about changes to the countryside (Young, 1988). Industrial pollution was identified as the greatest threat to the countryside by around a third of respondents (32 per cent), followed by chemicals and pesticides (18 per cent), urban growth (16 per cent), motorways and roads (11 per cent) and the removal of the landscape by agriculture (11 per cent) (*Ibid.*).

Whilst agricultural practices generated the greatest level of concern, the Social Attitudes Survey also revealed a hardening of opinion against the development of urban and industrial land uses in the countryside (Table 7.1). In political terms this latter trend was as significant as the growing scepticism towards productivist agriculture, which only gradually exerted an influence over agricultural policy. In contrast, anti-development sentiments found an immediate expression in local campaigns against specific development proposals. Thus, as noted in chapter one, Mormont (1987) in charting the emergence of 'rural struggles' or rural conflicts, observed that 'the structural characteristics of the development of rural space hide a series of economic, social and political contradictions which make up the basis of "rural" forms of opposition' and suggested that the new 'rural question' concerned 'the specific functions of rural space and the type of development to encourage within it' (Mormont, 1987, p. 562).

Table 7.1 Public attitudes to development in the British countryside

	% agreeing				
	1984	1986	1987	1989	1990
Industry should be prevented from causing damage to the countryside even if this sometimes leads to higher prices	77	82	83	88	89
The countryside should be protected from development, even if this sometimes leads to fewer jobs	n/a	n/a	60	72	n/a

n/a not available

Source: British Social Attitudes Survey.

Later surveys, although less comprehensive in their questioning, indicate that the strengthening of public opinion against developments in the countryside continued into the 1990s. Support for the relaxation of planning laws to enable more people to live in the countryside, for example, fell from 34 per cent in 1987 to 23 per cent in 1999 (Park *et al.*, 2001). Yet, as Lowe *et al.* (1993) note, the British Social Attitude Survey also identifies regional differences in attitudes towards development in the countryside. The 1987 survey found that a majority of respondents in Scotland placed the creation of new jobs ahead of countryside protection, whilst in the south east of England, countryside protection was favoured by 68 per cent to 26 per cent (Young, 1988). Moreover, countryside protection was placed ahead of job creation by respondents who lived in the countryside by a majority of three to one. Hence, the evident conservationist majority cannot be attributed to ill-founded urban opinion, but rather seems to represent a reactionary stance by rural residents seeking to protect their investment or their undeveloped rural locality. Nonetheless, the findings of the Social Attitudes Survey put public opinion at variance with the discursive assumptions that still underpinned much of government policy towards rural planning and the priorities of rural political elites.

Contested Discourses of Development in the Countryside

For much of the 20th century, a pro-development coalition was dominant within rural policy making. Nationally, policy was guided by a utilitarian discourse of the rural as a less populated space in which large or noxious land uses could be developed more cheaply and with less disruption than in more urbanized regions. As such, rural sites were developed as the locations of reservoirs, power stations (both fossil fuel-burning and nuclear), mineral workings, waste incinerators and landfills, and airports. Rural space was traversed by new roads connecting urban centres, and accommodated the over-spill of new housing from urban areas, either through incremental suburban expansion or through the construction of new towns on greenfield sites. Industrial development was also actively encouraged in many rural localities, with sites often marketed as providing space for expansion to foot-loose companies looking to escape from constrained urban premises. Developments of this kind therefore not only serviced an urban-based population and economy, but were also perceived to address the key problems facing rural regions – peripherality, lack of infrastructure and a shortage of jobs. The development of the countryside, it was believed, brought both infrastructural improvements and new jobs, thus boosting lagging rural economies. As such, the pro-development agenda also tended to be supported and promoted by local rural elites, so long as their own economic interests were not threatened. Generally, this meant that the interests of agricultural capital were protected, an objective that was achieved through the ordering of rural space, with top grade agricultural land excluded from development sites. Similarly, spatial ordering was employed to address the concerns of the rural preservationist movement (Murdoch and Lowe, 2003). National parks, greenbelts and sites of special scientific interests were designated to protect the most cherished landscapes and habitats and to contain urban sprawl, whilst sacrificing other parts of the countryside for development. This strategy reflected an understanding of nature as being both wild and resilient (see also Frouws, 1998):

From this perspective, the wilderness of rural landscape requires taming through road-building, bridge-building, electrification and so on to make it hospitable for human activity; whilst at the same time, the rural landscape offers the opportunity for the harnessing of *natural resources* for human service – through mining and quarrying, forestry and agriculture, the creation of reservoirs and the generation of hydro- and wind- power. Nature as resilient can withstand such human intervention, such that developments are perceived neither as unnatural nor as threats to the long-term survival of nature. (Woods, 2003b, p. 273).

Public opinion in Britain, however, appears increasingly to be following an alternative discourse, in which nature is conceived of as pure, idyllic and vulnerable, and rural landscapes are positioned as requiring protection from harmful human intervention. The modification of the rural landscape by human agency is recognized, but is only considered to be acceptable within certain parameters. Thus, human artefacts are acceptable if they are essentially biological (such as crops and forestry), or if they employ local natural resources in small-scale developments that conform to the prevailing natural aesthetic of the landscape (a dry stone wall, for instance, or an isolated farmbuilding). In contrast, developments that introduce large quantities of 'alien' material, such as tarmac or metal, or modern technology, or that appear disproportionate in scale to the morphology of the landscape, or that introduce noise or light pollution, are considered to be unnatural and 'out of place' (Woods, 2003b). Whereas the utilitarian discourse has provided the driving force for the development of the countryside, this latter discourse has set the limits of public tolerance. As such a number of thresholds have been passed over the last 40 years at which various types of development have become regarded as unacceptable. The submersion of significant areas of rural land to create water-storage reservoirs serving urban centres, for example, passed this threshold in the 1960s following protests against new reservoirs in Upper Teesdale, on Dartmoor and at Tryweryn, Clywedog and Craig Goch in Wales (Bull *et al.*, 1984). As later sections of this chapter will illustrate, major road-building in rural areas passed the threshold of unacceptability in the mid 1990s, and campaigns in the late 1990s showed unchecked housing development to be unacceptable. In the first decade of the twenty-first century, a new threshold appears to be being approached over the acceptability of wind power stations in the rural landscape.

The equilibrium between these two perspectives has been formally maintained through the town and country planning system in Britain, which has in theory sought to plan for development whilst protecting the environment. However, as Cloke and Little (1990) demonstrated, the planning system is part of the capitalist state and as such reflects the interests of economic capital that are generally best served through the development of rural space (Cloke and Little note that in certain localities, the planning system can also act as an expression of class conflict in restricting development to protect the property investment of the middle class – which is discussed later in this chapter):

Rural planning and policies can also be seen as part of the state's apparatus of *production*. Indeed, there are many policy mechanisms which have contributed towards the creation and maintenance of an environment for capital accumulation in rural localities. The underpinning of agriculture through policies of subsidy and land-use planning controls; the attempts to ensure a continuing supply for land for housing construction; the interpretation of planning regulations to permit industrial and commercial activities on green-field sites; and the relaxation of planning opposition to major developments in rural areas such as nuclear power stations, mineral workings, and transportation routes are all examples of the apparatus of production in rural areas. (Cloke and Little, 1990, pp. 59-60, italics in the original).

This bias is manifested, for example, in the terms of engagement of the planning system. Objections to planning applications can only be successfully pursued if they are expressed in the correct technical language and are based on legally-defined appropriate grounds for rejection. The planning system is framed to work with quantifiable (or at least, estimatable), 'objective' evidence, and allows little space for emotional arguments. In this way, campaigners whose opposition to proposed developments in rural space is motivated by concern about the impact of the amorphous 'rural idyll', about the disruption of a view, or about the lowering of property values, have frequently found that the planning system has proved unresponsive to their representations. As such, conflicts about developments in rural areas have generated public protests and media campaigns as well as lobbying through the planning system. This chapter examines some of these protests and campaigns by focusing on a key area of conflict, housing development, before briefly discussing conflict over road building and the construction of wind power stations. First, however, it establishes the context by examining in more detail the trajectories of the pro-development and preservationist discourses in rural politics.

Rural Modernization

In 1851 half of the population of England and Wales lived in rural areas. A century later, in 1951, only one fifth of the population did so. During that century, the total population of England and Wales had increased by 26 million people (or 144 per cent), yet the population of rural areas had *fallen* by some half-a-million people (or a 5 per cent decrease) (Saville, 1957). It was this context of rural depopulation that shaped attitudes towards development in the countryside for much of the twentieth century, articulated through a discourse of modernization. The problem faced by rural areas, it was surmised, was that they lagged behind cities in their level of technological advancement, the job opportunities offered and the living standards of residents. In order to keep people in the countryside (which was desirable both for reasons of the management of the national population and to provide labour for rural capitalism), the modernization of the rural economy, rural infrastructure and rural housing were all believed to be necessary.

The objective of rural modernization formed the rationale for the establishment of the Rural Development Commission in 1910 with a remit of supporting small scale industrial development and improvements to infrastructure. It was also eagerly taken up by the new institutions of local government in rural areas. A volume published to mark the fiftieth anniversary of county councils in 1939, for example, notes that the most popular committee among county councillors was the committee dealing with highways and bridges, and both the national and county-specific sections of the volume boast photographs and maps of new road schemes and bridges constructed by county councils (County Councils Association, 1939). Equally, the volume points to the role played by county councils in refurbishing rural housing and in building hospitals, schools and mental health institutions, many of which were located in rural sites. Rural district councils also discharged a duty to modernize rural housing, constructing enclaves of council houses in villages and hamlets. This activity, however, was often tempered by the interests of local agrarian elites in controlling the supply of public housing in order not to undermine the paternalistic benefits of tied housing provision (Newby et al., 1978), and the rural class structure was frequently imprinted on to the landscape by the physical separation of the new council houses from established villages. Similarly, many of the urban district councils and borough councils that governed rural market towns pursued expansionist policies, building housing estates and industrial estates, and creating reservoirs in their rural hinterlands to supply municipal water systems.

The implementation of such policies, however, was characterized by geographical differentiation. In relatively prosperous agricultural regions, such as East Anglia, local elites representing agrarian capital interests moderated the extent of developments that could potentially threaten their power base (see Newby et al., 1978). In more peripheral regions, with more marginal conditions for agriculture, the discourse of rural modernization was enthusiastically embraced by both state agencies and local political elites, as can be seen in the example of mid Wales. The mid Wales region, comprising the pre-1974 counties of Cardiganshire, Merionethshire, Montgomeryshire, Radnorshire and most of Brecknockshire, experienced the most severe depopulation of any part of England and Wales, losing just under a quarter of its population between 1901 and 1961. A report on the problem published in 1964 concluded that the decline in population was connected to the poor infrastructure of the region and the limited employment opportunities, but identified prospects for regeneration through agricultural modernization, afforestation, the exploitation of natural resources and industrial development (Welsh Office, 1964).

This strategy for developing mid Wales was most forcibly articulated in a Liberal party pamphlet, co-authored by the then MP for Montgomeryshire, Emlyn Hooson (Hooson and Jenkins, 1965). In a clear expression of the pro-development discourse, the pamphlet called for the creation of a Rural Development Corporation and outlined a 'plan for mid Wales' that heavily relied on large-scale infrastructure and industrial development. Hooson and Jenkins proposed the expansion of Aberystwyth through industrial development towards a target population of 60,000 by 1999; with the concentration of the remaining regional population in satellite towns and village clusters, served by light industries that processed local resources. Other proposals included the construction of a 'first class' airport at Aberystwyth, a major tourist complex on Lake Bala, and the building of a motorway between Shrewsbury and Aberystwyth, a North-South Wales trunk road, and a coastal toll road. It is a telling reflection of the pro-development discourse that the coastal highway was a perceived as a way of enhancing the enjoyment of the environment and boosting tourism:

> We anticipate that a panoramic coast road would be built down the whole of the West coast of Wales. Bridges, or alternatively, barrages should be built to carry this road across the Mawddach and Dovey estuaries... Clearly it would have immense attractions to holiday-makers, for it would open up the wonderful panoramic views along this coast. If barrages should prove feasible, they would themselves add to the attraction of the area for boating and sailing. They might also provide fresh water reservoirs. (Hooson and Jenkins, 1965, p. 17).

Although this ambitious plan was never adopted, some of the ideas and principles contained within it were implemented in a more modest form by the Development Board for Rural Wales established in 1977.

As Madgwick's research on Cardiganshire politics in the early 1970s demonstrates, the notion that industrialization and new road building represented the hope for the future of rural mid Wales was broadly supported by local political elites (Madgwick et al., 1973). Yet Madgwick also recorded scepticism about the likely success of initiatives to develop industry or tourism, as well concerns for the impact on the Welsh language and a 'feeling that Cardiganshire ought to be preserved as it is today' (*Ibid.*, p. 176). The significance of the preservationist position in the region increased during the 1980s and 1990s with the growth in environmental awareness, supported in part by counterurbanization and in particular the in-migration of 'back-to-the-land' settlers. However, the agrarian and business elites, who are still influential in local government in rural Wales, have remained largely favourable towards the development discourse, periodically lobbying for new roads or for major developments such as prisons as centre-pieces of their vision for economic development in the region.

At a national scale, the significance of the pro-development discourse has weakened such that its main advocates are now industry groups, such as the Housebuilders' Federation or the nuclear power industry, who have a vested interest in development projects and whose arguments are framed in economic rationalist terms. They are supported by right-wing libertarians, for whom the regulation of development in rural areas is bureaucratic 'red tape' that actually harms rural communities. For example, a volume of essays produced by the Social Affairs Unit in the late 1990s argued for the relaxation of planning controls and the abolition of greenbelts, as well as for new roads, which it claimed were essential to rural life (Mosbacher and Anderson, 1999). Yet, such opinions are now firmly the minority position, replaced in the mainstream of rural policy by an anti-development discourse that has evolved from an alternative preservationist tradition.

The Preservationist Tradition

The preservationist movement in Britain has its origins in eighteenth and early nineteenth century romanticism, as popularized by writers such as Wordsworth and William Morris, and in the growth of interest in natural history and nature study in the mid nineteenth century (Bunce, 1994). As these fields developed in tandem with the industrial revolution and the great age of urbanization, so the cultural celebration of the rural landscape and of nature found occasional political expression in concerns about the loss of amenity sites to urban expansion or agricultural modernization. By and large, however, the early preservationist movement was focused on the protection of specific pieces of land that were imbued with particular natural or amenity value. The Commons Preservation Society, founded in 1865 and described as the 'world's first private environmental group' (Sutton, 2000, p. 89), campaigned to protect open spaces on the edge of cities for amenity use. The National Trust for Places of Historic Interest or Natural Beauty, established in 1895, sought to acquire and manage threatened land and buildings for the national benefit; whilst the Society for the Promotion of Nature Reserves, formed in 1912, aimed to persuade the National Trust and private landowners to set aside land as reserves for the protection of wildlife and flora (Bunce, 1994).

In the wake of the First World War, however, the scope of the preservationist movement broadened to encompass concerns about the future countryside as a whole. This reflected a general perception that the war years had marked a watershed in the advance of urbanization in Britain, putting the very character of rural Britain – and particularly, rural England – in danger. One famous cartoon, published in *Punch* magazine, depicted 'Mr William Smith' leaving a bucolic country village for military service in 1914, and returning to an industrial urban sprawl in 1919. Notably, the cartoon was reproduced by Clough Williams-Ellis as the frontispiece to his influential and polemical book, *England and the Octopus*, published in 1928. In his text, Williams-Ellis articulated the preservationist discourse of rural England as a precious yet fragile entity that was being strangled by the unregulated advance of urbanization, industrialization and commercialism. This analysis had a strong resonance with the rhetoric of inter-war Conservatism, promoted by figures such as Stanley Baldwin (see chapter four). In both cases, the vision of the threatened English countryside was presented in strongly nationalistic terms (see also Matless, 1990, 1998). Williams-Ellis, like the *Punch* cartoonist, believed that the advance of urbanization and industrialization represented a betrayal of the soldiers who fought to defend England in the First World War:

> In the late War we were invited to fight to preserve England. We believed, we fought. It may be well to preserve England, but better to have an England worth preserving. We saved our country that we might ourselves destroy it. (Williams-Ellis, 1928, p. 20).

However, as Matless (1998) observes, 'if there is any political philosophy out of tune with preservationist planning, it is Baldwinite Conservatism' (p. 30). Baldwin's timeless countryside was a rhetorical device, aimed primarily at protecting the power and influence of the landed class and the new agrarian elite. It was fundamentally an apolitical space (see chapters one and four), which could not accommodate the radical demands of the preservationists. As such, Williams-Ellis dismissed the Conservative party as seeking 'to conserve the wrong things' (1928, p. 73), and whilst the preservationist movement drew support from the ranks of the aristocracy and landed gentry, Williams-Ellis attacked landowners who encouraged development as a means of easing their financial pressures. In a fictional exchange, the charge is put to 'Mr. Otherman' – 'a substantial citizen' – by an accusing angel:

> Having so signally failed in your role of Country Gentleman, and having taken your mean revenge on all that had supported your empty claim to pretension, you bought a slice of Downland on the coast, where you built yourself an appropriate villa that hallooed its challenge in a strident Cockney voice over three miles of rolling downs, the sheer white cliffs and the defenceless sea. From this decoy house you looked upon your land and saw that it was good – good for exploitation – 'Ripe for development'... You found five thousand acres of Downland pasture – the immemorial resort of peaceful wanderers from the adjacent towns: you left it a forlorn and straggling camp of slatternly shacks and gimcrack bungalows, unfinished roads and weather-beaten advertisements. You brought chaos out of order, you created Golgotha-on-Sea, you acted blasphemously, and the world and your fellows are the worse for your living. (Williams-Ellis, 1928, pp. 64, 66).

Neither, though, was the preservationist discourse of rurality the same as a modern environmentalist perspective. Williams-Ellis and his allies were concerned not with damage to the environment, but with the erosion of the distinctive 'character' of the English countryside. Indeed, in *England and the Octopus*, Williams-Ellis discusses at length changes to the urban landscape, but it is the confusion of the urban and the rural that draws particular ire (see also Murdoch and Lowe, 2003):

> We plant trees in the town and bungalows in the country, thus averaging England out into a dull uneventfulness whereby one place becomes much the same as any other – all incentives to exploration being thus removed at the same time as the great network of smoothed-out concrete roads is completed. (Williams-Ellis, 1928, p. 21).

Thus, as is discussed further below, the preservationist movement was not anti-development *per se*, but believed that development needed to be properly planned, taking account of local character, the natural aesthetics of the landscape and respecting a spatial differentiation of the city and the countryside. Whilst Williams-Ellis suggested that such an agenda could be carried into parliament and local government by the entryism of preservationist candidates, the movement instead found its political voice and influence in the recently formed pressure group, the Council for the Preservation of Rural England (CPRE).

The CPRE

The Council for the Preservation of Rural England (CPRE) was founded in 1926 by three individuals with prominent positions within professions directly concerned with planning and development: the Earl of Crawford and Balcarres, chairman of the Fine Arts Commission; Guy Dauber, president of the Royal Institute of British Architects; and, most influentially, Professor Patrick Abercrombie, the president of the Town Planning Institute (Murdoch and Lowe, 2003). Other early members included civil servants, local government officers, academics, designers and activists in other preservationist groups. As such, the CPRE aimed to create a network of key professional individuals and organizations committed to countryside preservation. The agenda pursued by the CPRE was a technocratic one from the start, promoting the virtues of planning and good design, and reflecting a faith in new professions such as town planning and (relatively) new institutions such as county councils to make difference. Yet, this technocratic approach was combined with a nationalistic, idealistic, discourse of rurality. In proposing the foundation of the organization, Abercrombie conjured up a bucolic vision of the countryside, which he described as the essence of England:

> The greatest historical monument that we possess, the most essential thing which is England, is the Countryside, the Market Town, the Village, the Hedgerow Trees, the Lanes, the Copses, the Streams and the Farmsteads. (quoted by Lowe and Goyder, 1983, p. 18).

As Murdoch and Lowe (2003) record, the aim of the CPRE was described by one early senior member as being to encourage, 'all things of true value and beauty, and the scientific and orderly development of local resources' (p. 321). The rationale of the society was therefore not to oppose development *per se*, but rather to advocate planned and orderly development. New roads, new houses, industrial and commercial developments, roadside service stations and tea rooms, advertising and highway signs and electricity pylons could all be accommodated in the English countryside, if built in the right place and if designed in sympathy with their surroundings (see Matless, 1990, 1998). Indeed, preservationist writers such as Noel Carrington and Thomas Sharp often indicated a preference for the more land-hungry, dual carriageway style of highway, cutting larger arcs around settlements, over single carriageway arterial roads and short by-passes because the former were perceived to have greater aesthetic value (Carrington, 1939; Sharp, 1946; see also Matless 1994, 1998). Crucial to this approach was the idea of maintaining a sharp demarcation between urban and rural areas (Murdoch and Lowe, 2003). In language and imagery reminiscent of the notions of a natural order and of rural purity associated with the discourses of the 'country gentleman' and the agrarian countryside (see chapters 2 and 3), the rural-urban differentiation advanced by the preservationists implied a distinctive moral geography:

> Settlements are to have a form fit for their function; a town should be clearly and distinctly a town, a village a village. A normative geography of distinct urbanity and

rurality is asserted over an England-in-between of suburb, plotland and ribbon development. Such sites become anti-settlements, places not fit for any purpose, lacking visual and social composition. The aim is to produce a proper English settlement. (Matless, 1998, p. 32).

With a leadership drawn from the technocratic elite and sections of the landed establishment, the CPRE was well placed to garner political support for its campaign. Its inaugural meeting was addressed by the minister with responsibility for planning, Neville Chamberlain, who indicated an affinity with the society's aims by calling for concerted action to protect the English countryside, yet it was with the advent of a Labour government led by open-air enthusiast Ramsay MacDonald that the CPRE saw the real opportunity for legislation (Matless, 1998; Murdoch and Lowe, 2003). Lobbying by the CPRE helped to secure the passage of the Town and Country Planning Act of 1932 and the Restriction of Ribbon Development Act of 1935, although both measures fell short of the society's ambitions (Bunce, 1994). The issue of planning was nonetheless positioned on the political agenda and CPRE leaders were able to play an active part in a series of commissions that were established to consider the problems of land use and spatial planning from the late 1930s. Abercrombie served on the Royal Commission on the Distribution of the Industrial Population, which reported in 1940, whilst the influential Committee for Land Utilization in Rural Area had CPRE members as its chair (Lord Justice Scott) and vice-chair (L. Dudley Stamp) (Murdoch and Lowe, 2003). These commissions provided the groundwork for the establishment of the planning system by the Attlee Government, such that as Murdoch and Lowe observe, 'CPRE thinking thus set the agenda for post-war legislation, most notably the 1947 Town and Country Planning Act' (2003, p. 322).

The early successes of the CPRE had a number of consequences for the subsequent evolution and operation of the organization. Firstly, the CPRE had a vested interest in making the town and country planning system work. Lowe and Goyder (1983) argue that the society changed its style after the Second World War, from a 'promotional group', campaigning for a value change, to an 'emphasis group', not seeking to confront popular values but rather exercising vigilance on behalf of the preservation of the countryside by monitoring the operation of the planning system and engaging in its functions where appropriate. Much of this role was fulfilled by the CPRE's county branches, 30 of which had been established by 1939 increasing to over 40 by 2000. The branches enjoyed good relations with local planning authorities; in many counties the chief planning officer had a place on the CPRE branch executive and branches were frequently represented on county council countryside committees and on national park authorities (Lowe and Goyder, 1983). CPRE branches were routinely notified of planning applications and fed into the planning process both formally and informally. Through the volunteer labour of members, branches monitored planning decisions, responded to planning applications and made representations to public inquiries on structure plans. As such, the politics of rural preservation became largely channeled into the formal (and frequently, back-stage) processes of local government. This was replicated at national level, where the relatively small permanent staff of the CPRE (fewer than 10 during the 1960s) monitored and fed into national planning policy through formal and informal contacts with government departments.

Secondly, therefore, the influence of the CPRE during the post-war period rested on its technical expertise, its careful preparation and vigilance, and its ability to play the system. Winter (1996a), thus describes the CPRE as a 'potential insider' or 'thresholder' group, not being as tightly drawn into policy-making as groups such as the NFU (see chapter six), but enjoying privileged access to policy-makers whilst maintaining a degree of independence. The close relationship between the CPRE and both national and local government has become both the society's strength and its weakness. On the one hand, the organization's opinion has credibility in government circles:

> Civil servants know that they can come to us and they will get very thoughtful, logical, rational, credible policies, and we have a very strong reputation based on ensuring that CPRE publications and our policy process is very robust. Unlike some of the other voluntary organizations who engage in rural public policy we test most of our approaches on our policy committee which is made up of experts: chairs at various universities, local government people, retired civil servants. (CPRE chief executive, interview, January 2002).

Yet, on the other hand, in order to maintain good relations with government the CPRE was forced on occasion to moderate its position and accept compromises. The style of Sir Herbert Griffin, the general secretary of the CPRE from 1926 to 1965, for example, was described by one colleague as being 'personal contact in the corridors of power. He fastidiously avoided embarrassing those whom he influenced or sought to influence' (quoted by Lowe and Goyder, 1983, p. 75). Similarly, county branches carefully avoided conflict and confrontation with local government:

> An established group will also weigh carefully the consequences to its future relations with the local authority of adopting a stance of public opposition. In the words of the secretary of the Lancashire CPRE: 'Fighting cases means publicity. Publicity means conflict, and conflict can mean loss of contact and credibility'. (Lowe and Goyder, 1983, p 94 (after Allison, 1975)).

Thirdly, the CPRE was obliged to confront what Murdoch and Lowe (2003) describe as the 'preservationist paradox': that the division between urban and rural established in policy was immediately transgressed in practice, most notably by counterurbanization. In seeking to preserve the countryside, the principles of spatial differentiation in the planning system actually encouraged urban to rural migration:

> By protecting rural areas, planning simply makes them more attractive to urban migrants, especially in a context of car-based mobility which allowed individuals and households to move in ever greater numbers across urban and rural spaces. (Murdoch and Lowe, 2003, p. 323).

Moreover, many in-migrants to rural areas became active supporters of the CPRE, helping to swell the society's membership from 15,000 in the 1960s to around 40,000 in the 1990s (Lowe *et al.*, 2001), as they mobilized to defend their financial and emotional investment in their new homes. The growth in membership changed the character of the CPRE, away from its elitist roots. Whereas in the early 1970s, the chairs and secretaries of CPRE branches were often established rural residents of 'high social standing', including members of the House of Lords, magistrates and members of historically important families (Allison, 1975), by the end of the same decade the leadership of local branches was beginning to pass to in-migrant professionals. A survey of committee members of the Suffolk Preservation Society (the local branch of the CPRE) in 1977, for example, found that the traditional county establishment still had a presence – the society's patron was the Earl of Stradbroke, its president the Duke of Grafton, and there were a further three peers, three magistrates and five holders of military titles on its committees – but that more than half of the committee members were 'professional people or retired professionals – teachers, planners, surveyors, architects or doctors' (Buller and Lowe, 1982, p. 24). A third of committee members were described as 'urbanites' who had either retired to Suffolk (23 per cent), or commuted out of the county to work (10 per cent); whilst a further quarter were local professionals and businesspeople who had moved to Suffolk within the previous twenty years. Only 7 per cent were farmers or landowners (Buller and Lowe, 1982).

The tone and focus of the CPRE's activities also changed under pressure from two directions. Many of its new members held more uncompromising views on development in countryside than the CPRE's conventional position. They often tended to oppose new developments outright, rather than the conventional support for planned development, and many new members believed that the CPRE should adopt a higher campaigning profile. At the same time, the emergence of a broad-based environmental movement, which also opposed particular developments in rural space on ecological grounds, forced the CPRE to rethink its preservationism within the new environmental agenda. Thus, Murdoch and Lowe (2003) argue that the CPRE adopted an environmental approach, positioning countryside preservation within a 'ecological' frame of reference which highlighted the relational connections of rural change with wider social, economic and environmental factors. An early expression of the shift in emphasis was a change in the organization's name to the Council for the *Protection* of Rural England in 1969, which was followed by a further change to the *Campaign to Protect* Rural England in 2003.

A further dimension was the strengthening of the CPRE's headquarters operation, with an increase in the number of full-time staff from 13 in 1980 to 47 in 2004, including a 14-strong policy and campaigns unit, plus 12 regional policy officers. The six communications staff included in the headquarters total reflect the importance now placed by the CPRE on getting its message across to the public, also reflected in the appointment of media figures such as Jonathan Dimbleby and Sir Max Hastings as the society's president as opposed to the aristocrats and politicians of the past. Thus, as Murdoch and Lowe observe, the headquarters 'is charged with more effectively communicating the CPRE perspective to policymakers, the media and the branches. In effect, the central office works to set local preservationism within a context of national environmentalism' (2003, p. 324).

The main focus of the CPRE, however, continues to be the development of informed, coherent, critiques of policy and formal engagement with the policy process through consultation exercises and lobbying, albeit with one eye on the comparative profile of other rural lobbying groups:

> CPRE will always see the behind-the-scenes, influencing the green paper and the white paper, talking to civil servants about the consultation document, quietly behind the scenes, as more important than the blowing the trumpet once you've got the legislation and effectively its too late to do anything but it makes you look good and you get the profile. But what we do have to take account of, and increasingly so, is more competitors coming in. Members have to see that we are doing things, in relation to other organizations, like the National Trust, increasingly trying to boost their profile. So, the National Trust getting involved in policy issues, Countryside Alliance engaging other rural issues other than hunting. If people are going to carry on supporting CPRE, if volunteers are going to carry on giving effectively three days a week, they have to see what we are doing. You have to create news, but that will always be for us a secondary practice. (CPRE chief executive, interview, January 2002).

At the same time, the CPRE's approach has also been challenged by proposals to increase direct public participation in the planning process, opening the organization to charges of elitism:

> The government have just launched a planning green paper, which is fundamentally about a reform, taking forward to legislation, the way the planning system works. The CPRE is in quite a difficult position because there's a lot in the government's reforms that we're quite uncomfortable with, which we believe will fundamentally damage the ability to protect the environment and curtail democratic engagement, but civil servants say to us, 'well actually, what you're arguing about is you're your in favour of the status quo because you know how to engage, you're an intelligent, articulate organization with volunteers who engage on a one to one level, but what we want to do is to open the process up to individual members of the community, whereas what you want to do is carry on talking about structure plans, and three tiers in the planning process, because you know how to engage'. It's an interesting debate, because we in a sense are seen to be part of the problem, because there's a need for reform and we're not seen as an organization that's actually able to represent rural communities, we represent a distinct entity. (CPRE chief executive, interview, January 2002).

Such perceptions tend to structured around the activities of the CPRE at a national level. Locally, the branches of the CPRE can vary considerably in the scope of their activities, the interests that they represent and their relationship with local government. One recent study has shown that the compatibility between the CPRE's national and local activities is greatest in the south east of England, but that in more peripheral regions particular local factors can intercede (Lowe *et al.*, 2001). Thus, in Hertfordshire, the local branch replicates many of the characteristics of the CPRE nationally. With its outer metropolitan location, the membership of the branch is rich in higher order professional skills and expertise – at the time of the research, the branch's planning manager was a retired civil servant from the Department of the Environment, and the chair of its planning policy committee was a retired planning solicitor. Branch officers identified a strong correlation between their work and that of the national CPRE, and the branch enjoyed a good relationship with local government, with whom they shared a broad consensus on planning policy. In contrast, the Devon branch had grown with the recruitment of in-migrants during the 1980s, many of whom were motivated by primarily local concerns and felt that material from head office exhibited a 'home counties bias'. The Devon branch also had a more fractious relationship with local government, reflecting the local political context. Whilst planners respected the professionalism of the CPRE, councillors drawn from the established rural elites dismissed the society as 'middle class incomers', 'narrow and anti-development', or, in the phrase of one councillor, 'the Council for the Ossification of Rural England' (Lowe *et al.*, 2001). Finally, in Northumberland the absence of a significant rural middle class had restricted the development of the local CPRE branch, which in consequence was fairly weak in its local political significance and heavily reliant on the national and regional offices of the CPRE (Lowe *et al.*, 2001).

These local geographical factors became increasing significant for the CPRE's activities, compared with those of other rural and environment groups and ad hoc campaigns, as the scale of housing development in the countryside became a major political issue during the 1990s.

Housing Development in the Countryside

The trend of counterurbanization in the second half of the twentieth century has had a major impact on the landscape and environment of rural Britain. As noted in chapter one, the population of rural counties in Britain increased by six per cent between 1971 and 1981, and the population of rural districts in England increased by a further 12.4 per cent between 1981 and 2001. At first, in-migrants to rural areas tended to purchase existing housing stock that had become vacant through the consequences of social and economic change – from old manor houses sold off by the dwindling squirearchy to farmhouses which had lost their land in farm amalgamations to renovated farmworkers' cottages. The tide of counterurbanization was such, however, that demand for rural properties soon outstripped demand, contributing to an escalation in house prices. At the same time, counterurbanizers were being drawn from an increasing broad range of social backgrounds, including many who aspired to a rural lifestyle but required the availability of relatively cheap housing in order to be able to relocate. The solution was to build new housing in the countryside, as the editor of *Country Life* magazine cynically observed,

The trouble is that the New Countryman wants quality of life but does not want to pay for it. There are plenty of houses already in the countryside, but they have become expensive. Too many people are after them. Now the answer might seem to be to build more houses. That is what has happened over the past decade. (Aslet, 1991, p. 42).

According to the CPRE, a total of 705,000 hectares changed from rural to urban land uses in England between 1945 and 1990 – or the equivalent of urbanizing an area the size of Greater London each decade (Sinclair, 1992). Whilst this figure also includes land developed for industry and commerce, much of it can be attributed to new housing development and the associated infrastructure. As well as the pressure from counterurbanization, the demand for new housing in rural areas has come from factors such as the changing structure of households, the greater longevity of the population, and the continuing flow of population from remoter rural settlements into small towns and larger villages.

The development of new housing has been regulated through the planning system, following the principle of spatial differentiation described above. Rural land around major conurbations has been protected by the designation of 'greenbelts' within which new development has been virtually prohibited. Although greenbelts have been successful in checking urban sprawl, they have proved less successful in protecting the countryside as a whole. Migrants from cities have simply leap-frogged the greenbelts, placing more demand on the rural districts that lie immediately beyond them. Within the broader countryside the principle of spatial differentiation has tended to be followed through the use of 'key settlement' policies (Cloke, 1983). New development was concentrated on the fringes of medium-sized towns without greenbelt restraints, in market towns and in larger villages that were considered to be capable of absorbing expansion. New developments in the open countryside were essentially forbidden, and smaller, more upmarket, villages were usually protected from significant new development. For the most part this approach worked smoothly, striking a balance between the demands for development and preservationist concerns. Occasional conflicts were sparked by specific housing developments, and lobby groups such as the CPRE and the Housebuilders' Federation competed to inform structure plans, but the outcomes were broadly accepted.

In the late 1980s and early 1990s, however, opposition to new housing developments in rural areas began to strengthen. At a local level, new in-migrants became more bullish about proposed developments in their adopted communities, particularly in 'key settlements' whose expansion had facilitated mass counterurbanization but which were now perceived as being in danger of becoming 'too urban' in character. At a national level, meanwhile, concerns also developed about the cumulative impact of rapid housing development on the English countryside as a whole. In March 1995 the Department of the Environment published projections that the number of households in England would increase by 4.4 million between 1991 and 2016, largely as a result of changing patterns of household structure (Gilg, 1999; see also Gummer, 1998). Moreover, it was forecast that when migration flows and residential preferences were taken into account, around half of the new households would need to be accommodated by new build on greenfield sites in rural areas. Although the projection was based on a scientific analysis, and was considered by some commentators to be a low estimate of the likely demand (see Gummer, 1998), the imagery of over two million new houses being constructed in the English countryside was powerfully reproduced by campaigners against the scale of development. As such, a conflict over housing development was sparked that was played out at three levels: the national scale, contesting the overall target and policy; the county scale, as councils struggled to convert projected housing demands into structure plans; and at the local scale, as protests were mounted against individual development sites.

The three examples discussed below illustrate the articulation of the conflict over housing development at each of these scales. The first example, which draws on research by Murdoch and Abram (2002), discusses the contestation of proposed housing development at a village level in Buckinghamshire. The second example examines debates over the allocation of new housing at a district level in Taunton Deane, Somerset. Finally, the third example follows the emergence of a national debate about housing development in the countryside.

Village Preservation in Buckinghamshire

Buckinghamshire was the fastest growing county in England during the 1970s, with a 20 per cent increase in the number of households between 1971 and 1981, followed by a further 10 per cent increase between 1981 and 1991. The intense pace of growth was fuelled by out-migration from London, the promotion of Milton Keynes as a new town, and the prosperity of the south-east England economy. With the regional economy continuing to boom, government projections forecast that Buckinghamshire would have the second highest rate of growth between 1991 and 2016 (after Cambridgeshire), with a 29.3 per cent increase in households. However, the distribution of new housing development to accommodate the growth in population since 1970 has been spatially uneven. Reflecting government planning policy, there has been minimal development in the southern districts of Chiltern and South Buckinghamshire, which fall within the London greenbelt. Much of the growth had been absorbed by the new town of Milton Keynes, the expansion of which had appropriated significant areas of rural land in the north of the county, yet the district of Aylesbury Vale in the centre of the county had also experienced considerable development as an attractive rural area but without greenbelt restrictions (Abram et al., 1996; Murdoch and Marsden 1994).

The geography of housing development in Buckinghamshire was also influenced by local political movements. A very strong anti-growth coalition in the south of the county helped to restrict development in the area, whilst the greater pressure placed on Aylesbury Vale reflected the relative weakness of organized opposition of development in mid Buckinghamshire. At a local level, the restriction of development in individual communities was advanced by village amenity societies established in the 1960s and 1970s. Many of the village societies had strong links with the CPRE and followed a similar mode of operation, harnessing professional skills and working through engagement with the planning system. In doing so, they aimed to 'use the planning system to shape the village in accordance with ideal images of the rural community' (Abram et al., 1996, p. 361). Thus, the objectives of one village society were described as being 'to try to keep [the village] a pleasant place to live rather than be anything too left-wing or "green orientated"' (village society leader quoted by Abram et al., 1996, p. 361). The protection of the 'rural idyll' came under more pressure, however, in the 1990s as the process of reviewing the Structure Plan and the various District Plans commenced, particularly in Aylesbury Vale, which had been required to accommodate over a fifth of all new housing development in the county between 1991 and 2011 (Table 7.2).

Table 7.2 Proposed housing development in Buckinghamshire, 1991-2011

	Number of new households
Milton Keynes	37,200
Aylesbury Vale	14,300
Wycombe	6,700
South Buckinghamshire	2,500
Chiltern	1,900
Buckinghamshire total	62,600

Source: Murdoch and Marsden (1994).

The proposed demand for new housing in Aylesbury Vale provoked a political reaction in the district, not least because residents of villages that had expanded rapidly under previous rounds of development were mobilizing, motivated by concerns that the settlements had reached their development limits. The response was also shaped by local political factors. The allocation of new housing development had been agreed under a Conservative administration which lost control of Aylesbury Vale District Council in 1995. The new Liberal Democrat administration, which was more inclined against development, 'was faced with an intractable dilemma of making itself unpopular with constituents by finding locations for large amounts of new housing, or facing severe judicial problems if it chose to allocate smaller numbers of housing sites' (Murdoch and Abram, 2002, p. 113). Some councillors argued that the council should simply tell the Department for the Environment, Transport and the Regions (DETR) that the district 'cannot cope' with the allocation, but planning officers cautioned against this stance. Instead, the council sought to reduce the environmental impact of new development by using a criteria based on 'sustainability principles' to identify sites where new housing could be more sustainably accommodated.

One of the sites identified was in the village of Haddenham, 8 kilometres south west of Aylesbury, which was allocated 400 extra houses. Haddenham had expanded rapidly since the Second World War, with its population increasing from around 500 in the 1940s to around 5,000 in the 1990s. This growth had included new council housing, much of which was bought by tenants during the 1980s, and new estates of 'executive' housing. The village had hence become socially mixed, but residents still identified a 'village feel', which they feared would be lost in any further expansion.

The rapid development of the village in the 1960s had prompted the formation of a Haddenham Village Society (HVS), which adopted a typical preservationist approach. Its members included planners and architects resident in the village, and the society campaigned for 'more architecturally and socially sensitive forms of planning in the village' (Murdoch and Abram, 2002, p. 120). As such, the response of the HVS to the proposed District Plan was to stress 'that they were unhappy about accepting any new housing allocation, but that in view of the strategic pressures, they recognized that some housing was inevitable' (Murdoch and Abram, 2002, p. 121). The HVS stressed design and location considerations and argued for the number of new houses to be kept as low as possible. However, the pragmatic stance of the HVS was not acceptable to a more militant group of villagers opposed to any development, who formed an alternative protest group, the Haddenham Protection Society (HPS). As Murdoch and Abram observe:

> The HPS demarcated itself quite clearly from the HVS. The new society began enthusiastically to raise awareness in the village about the District Local Plan proposals: it organized a petition to the Council undertook a fund-raising campaign, distributed window-posters declaring 'No to 400 houses in Haddenham', and organized parties of protesters to attend council meetings, at first carrying placards saying 'AVDC, keep off the grass please', and later wearing tee-shirts decorated with a picture of a scarecrow, and marked 'Save Our Habitat – Haddenham Village'. (Murdoch and Abram, 2002, p. 121).

In common with many local, ad hoc, protest groups, the Haddenham Protection Society rejected conventional organizational forms, operating as a tight-knit network of activists and focusing on a single issue. This helped to maximize their impact as, again, did local party politics. The district councillor for Haddenham had initially backed the new housing development, reflecting the position of the HVS and of longer-established villagers concerned about the lack of affordable housing in the community. Under pressure from the HPS, however, and interested in raising his profile following his selection as a Liberal Democrat parliamentary candidate, the councillor mobilized support on the council for a revision to the District Plan which dispersed the new housing developments more widely.

The revised plan allocated 150 new houses to Haddenham, which was still unacceptable to the HPS. Moreover, with more villages now identified for development, the HPS was able to forge a coalition with other village protest groups. In doing so, it also professionalized its approach, moderating its position in the process. One strategy adopted was to find alternative brownfield sites on which the new housing could be located. Whilst many of the brownfield sites were not favoured by planners because poor public transport links made them less sustainable on the planners' criteria, they offered a way out for councillors. In consequence, the number of greenfield sites was reduced. For the HPS, however, the strategy proved to be counter-productive. Its coalition with other villages crumbled and as the new plan gained support, the HPS found itself unable to further oppose a revised allocation of 100 new houses to Haddenham.

Housing Development and Rurality in Somerset

In contrast to Buckinghamshire, there has not historically been a strong anti-development lobby in Somerset. Yet, although Somerset lies outside the pressurized south east, the county has experienced significant growth since the 1970s, with 24,000 new houses built between 1981 and 1989. Moreover, Somerset was forecast to be the tenth fasted growing county in England between 1991 and 2016, with a projected 58,000 new households, or a 22.5 per cent increase. Prior to 1990, the politics of housing development in Somerset had been largely controlled within the planning system. The Somerset Structure Plan adopted in the early 1980s, for example, had planned for 36,750 new houses between 1981 and 1996, yet this scale of development provoked little public protest. Rather, the negotiation of the structure plan, and its subsequent revisions, was undertaken through a public inquiry and consultation with interest groups including the CPRE, the Country Landowners Association, local civic societies and the Housebuilders' Federation. These organizations shared a broad consensus with councillors and planners on the need for housing development, and the emphasis of the CPRE's contribution was on modification of the proposals not opposition to development.

In the 1990s, however, this consensus was undermined as discussion of the new structure plan, covering development up to 2011, progressed. This process commenced in July 1991 with the publication of a draft regional strategy that projected a need for between 365,000 and 400,000 new houses in the south west region, including between 47,000 and 50,000 in Somerset. The final regional plan, published in March 1993, confirmed the allocation of 50,000 new houses to Somerset, which subsequently formed the basis of the draft structure plan produced in February 1995. During the early stages of this process, participation was essentially restricted to civil servants, planners, technical advisers, senior councillors and established interest groups such as the CPRE. Although some local authorities, including Somerset County Council, did question the scale of the forecast housing demand, their challenges were framed through critiques of the methodology employed and the assumptions used in calculating the projections, and thus remained within a technical and statistical discourse.

Only following the publication of the draft structure plan was wider public awareness of the proposed housing development stimulated and more interest groups and protest campaigns drawn into the debate. As participation broadened, so the discursive terrain of the debate also shifted. From being a technical issue of projecting household growth, the debate now became one about the impact of the proposed development on the rural identity of the county. This was articulated in a number of ways. Firstly, the construction of 50,000 new houses was presented as a threshold that would take Somerset beyond its capacity for development and irredeemably damage the rural character of the county (see also Woods, 1998a):

> The way of rural village life which has slowly evolved over the centuries is going to be wiped away at a stroke. (Chair, Somerset CPRE, quoted in the *Somerset County Gazette*, 3 February 1995).

> There are certainly fears from established Tauntonians that if you take things much further, villages which stood alone would just get sucked into the general morass of the town. (Local newspaper editor, interview, November 1995).

> [The] housing balance must be right to protect rural life. (District councillor in letter to the *Somerset County Gazette*, 24 March 1995).

By adopting this position, groups that had supported – or at least not objected to – previous development could justify their change of position by representing the new proposals as an expansion too far. Secondly, and similarly, the housing projections were presented as being self-defeating because the scale of development would urbanize Somerset to the point at which prospective in-migrants would be deterred:

> There seem to be two strategies – expanding major settlements, including Taunton, or building a little bit on existing settlements ... You can only carry on doing that for so long because after a while you reach a crunch point where that policy starts destroying the framework of the countryside people want to move into. (Director, Somerset Community Council, quoted in the *Somerset County Gazette*, 1 December 1995).

> Objectors, which include the Green Party, say such a massive building programme will destroy the very countryside and quality of life people come to Somerset to enjoy. (*Somerset County Gazette*, 31 March 1995).

> The regional planning directive is for 50,000 more houses in Somerset. It's not going to be the Somerset people want. (I412 - District councillor, interview, 1994).

Thirdly, the debate about the scale of proposed housing development at a county level was punctuated by localized protests against specific housing developments that were already underway. In each of these cases protesters sought to defend some sense of the 'rural idyll', emphasizing the impact of the new development on the landscape, wildlife and residents' privacy:

> We object to any development at all. We came here from an estate in Taunton for the views and the privacy and that didn't come cheap. (Resident opposing a Housing Association development, quoted in the *Somerset County Gazette*, 16 June 1995).

> People don't want any more housing, they want to keep it as a village. We would rather see sheep than houses, we need to preserve Mill Lane, it's a unique rural setting. (Resident opposing a private housing development, quoted in the *Somerset County Gazette*, 12 July 1996).

> We won't be seeing any kingfishers, herons and badgers round Sherford stream once the bulldozers arrive. (Letter to the *Somerset County Gazette*, 14 April 1995).

In the above quotes the protection of 'rurality' is implicitly elided with a defence of property investment. As such, the protests might be seen as part of a defence of 'middle class space', as discussed in more detail later. At the same time, however, the class recomposition of the countryside was presented as an argument against the 50,000 proposed new houses. From this perspective it was the type of housing being proposed as much as the amount that was the problem. The chair of the Somerset CPRE, for example, argued that the housing would be bought by 'people relocating from areas like the South East who will merely use villages as a convenient dormitory' (quoted by the *Somerset County Gazette*, 3 February 1995), whilst a Labour councillor claimed that no one was building houses for local people and suggested that, 'until we meet local planning needs there should be a moratorium on providing migration housing' (quoted by the *Somerset County Gazette*, 31 March 1995).

Not everyone, however, was opposed to the housing development. A minority body of opinion sought to defend the proposals for more housing, albeit more vociferously in private than in public. In line with the established rural modernization discourse, supporters of more housing represented opponents as being against 'progress' and argued that some development was necessary to keep villages alive:

> [Councillor X] seems to believe that her village should somehow be exempt of progress. (Letter to the *Somerset County Gazette*, 26 May 1995).

> I think that you've got to have a bit of development if you want your own people to benefit from them, otherwise you find yourself out of touch with the demands, possible demands, from other parts of the country (I190 – Former county councillor)

> I used to get very hot under the collar because they would never give us any planning ... They're very against ribbon development along the rhines, God knows why, I still haven't actually cracked that one. There are masses of plots, and I fought, we used to fight, tooth and nail when I first got on the council, [for] one or two handy plots which you would have thought would have been absolutely excellent to put houses and things on. Because we are desperate for new buildings in this village, we really are. (I120 – Former district councillor).

Tellingly, favourable opinions towards the plan tended to be expressed by former councillors rather than serving councillors, reflecting the marginalization of a once more prominent pro-development position. Yet it was these people who were more in tune with the rationality of the planning system than the anti-development protesters, a fact that became apparent as the campaign intensified.

In Taunton Deane, which had been allocated 12,600 new houses, opposition was organized by the Campaign Against Unwanted Development, launched in March 1995 by Friends of the Earth, the CPRE and the Green Party, which called for the housing allocation for Somerset to be cut by 10,000 homes. The campaign provided a focal point for a broader seam of unorganized public protest, including letters sent to local newspapers and ad hoc protests against specific housing developments advanced through petitions, demonstrations and the lobbying of councillors. Significantly, with council elections due in May 1995, the inaugural meeting of the Campaign Against Unwanted Development was addressed by all three major political parties, and housing development became a key issue in the elections. The two main parties in district, the Conservatives and the Liberal Democrats, were however placed in difficult positions by the debate. The Conservatives championed the anti-development cause, not least because of the perceived threat to middle class villages in a number of marginal Conservative wards, and sought to blame the Liberal Democrat-controlled county council. In doing so, though, they downplayed the traditionally strong ties between the local Conservative party and the building industry, and the fact that it had been a Conservative administration that had originally agreed the 50,000 house target for Somerset. The Liberal Democrats, meanwhile, were largely sympathetic to the anti-development campaigners, but were forced to defend a policy that was not of their making. As discussed in more detail elsewhere (see Woods, 1998a), the Liberal Democrat leadership on both Somerset County Council and Taunton Deane Borough Council hence attempted to separate their responsibility to implement the structure plan from their right to represent the interests of their constituents. Thus, the defence of the structure plan proposals was largely left to planning officers, whilst councillors continued to express their reservations about the proposals, blaming them on the (Conservative) central government.

The National Debate

It was not only in Somerset that local councillors sought to shift the blame for the proposed housing development to central government. Across England, councils under pressure from local campaigners but legally compelled to accommodate the housing demand projections into structure plans and district plans, sought confrontation with the national government as a means of displacing criticism. West Sussex County Council deliberately adopted a structure plan that accommodated only 37,900 new houses – 12,800 fewer than it had been allocated by the regional planning strategy. When the council's assessment that it could not accommodate more than the 37,900 houses was upheld by an independent panel, the council was ordered to incorporate the 12,800 additional houses into its plan by the Deputy Prime Minister, John Prescott, who had ministerial responsibility for planning. In response, the council went to the High Court to request a judicial review of the minister's decision. Challenges to the government's housing allocations were also mounted by Devon, Cheshire, Hampshire and Gloucestershire county councils.

Other councils opted to be provocative by breaching established planning conventions in identifying sites for new housing. Stevenage Borough Council in Hertfordshire controversially allocated 10,000 new houses to a site within the London greenbelt. Despite opposition from the CPRE, the plan was approved by the DETR, as were smaller incursions into greenbelt designations on the edges of Bradford and Newcastle-upon-Tyne. Berkshire, Cambridgeshire and Hampshire county councils, meanwhile, all revived earlier models of urbanization by proposing that some of the new housing should be concentrated in the development of entirely new towns. Pressure was also increased by the adoption of direct action tactics by campaigners against developments in locations including Hockley in Essex, Peacehaven in East Sussex, and on the Bradford urban fringe, and suggestions that radical environmental groups such as Earth First! might join protests. These local strands of opposition were coalesced in a national campaign run by the CPRE, including press adverts portraying billboards advertising the English countryside for sale.

The upscaling of the housing development debate from local government to the national political arena coincided with the election of the Labour government in 1997. Like many council administrations that had gained power in 1995 with structure plan reviews in mid-process, the Labour government found itself compelled to implement and defend targets for housing development that had been agreed by the previous government. Significantly, however, Labour had accepted the projection of 4.4 million new households in opposition, and had even cautioned that the target for brownfield development should not be used to justify 'town cramming' with high density estates (Gilg, 1999). Yet, Labour's credentials on countryside preservation were perceived to be weak and the anti-housing development campaign became conflated with the mobilization of the wider rural movement in the Countryside March of March 1998, presented as another example of Labour's 'urban bias'. Additionally, Friends of the Earth targeted a number of Labour MPs in rural constituencies who it claimed could lose their seats if plans for greenfield development were not resisted, some of whom in turn were openly critical of the government's position.

Labour's response to these challenges has been somewhat confused, but can be distilled into three main elements. Firstly, the government moved to increase the proportion of new housing to be built on brownfield sites, linked to a strategy for urban renewal. However, whilst this strategy appeased some campaigners, it was criticized by the building industry which argued that there was insufficient brownfield land to accommodate the stated proportion of development in many counties with the greatest projections of growth, including Somerset, Berkshire and Suffolk. Indeed, by 1998 the building industry was promoting its own projection that the real demand for new housing would be nearer to 5.5 million than 4.4 million in the period up to 2016. In particular, the National Housing Federation forecast that 1.4 million new houses would be required in the south east of England between 1996 and 2016 – significantly higher than the 914,000 target for the longer period of 1991 to 2016 worked towards by Serplan, the regional planning body comprised of local government representatives (Hetherington, 2000). The DETR itself revised the projected demand to 1.1 million houses, 100,000 more than the already controversial original projection. Thus, secondly, in accepting a higher revised projection for the south east region, the government also changed tack in proposing that the expansion should be concentrated in new town developments in four major 'growth areas' focused on the Thames gateway, south Cambridgeshire, Milton Keynes and Ashford – generating a new wave of opposition from residents in the areas affected (Campbell, 2004). Finally, whilst conceding some ground to anti-development campaigners, the government has on the whole attempted to defend the need for more housing development. It has accused opponents of political opportunism and has highlighted problems of the shortage of affordable housing and spiraling house prices. As such, it has explicitly focused attention on the middle class nature of much opposition to new housing, revealing an imperative that underlies much of the debate about developing the countryside.

Defending Middle Class Space

The politics of development in the countryside have always been infused with a strong dose of class conflict. From the control over new development exercised by middle class business and agrarian elites in local government to protect their property interests and political influence (Newby *et al.*, 1978), to the technocratic tendencies of the preservationist movement, the rural middle classes have long ensured that their interests have been defended in any consideration of new development. Their ability to do so has been favoured by biases within the development control system, which mean that successful engagement requires both time and technical expertise. This is not to say that all opponents of development in rural communities are middle class, or that the rural working class is by implication pro-development, rather it is to recognize that the middle classes have been particularly active in conflicts over development in rural communities and that the politics of development in the countryside have increasingly become framed around the consequences of middle class investment in the countryside under counterurbanization. The investment of the middle classes has taken two forms. Firstly, there is an obvious financial investment involved in purchasing property. Secondly, however, for many middle class in-migrants there is also an emotional investment involved in cutting social and family ties and moving to a new location, often in pursuit of some idea of the 'rural idyll'. Having made this investment, it is understandable that middle class in-migrants are prepared to mobilize against perceived threats to the particular rural landscape, or rural community, or rural lifestyle that they have bought into.

In the pressurized countryside of south eastern England, the activism of middle class residents over a number of decades has succeeded in regulating development to the extent that many villages have become almost exclusively middle class communities in which a particular 'rural lifestyle' can be consumed within commuting distance of the urban centres on whose economic activity to maintenance of such a lifestyle depends. Murdoch and Marsden (1994) describe this process at work in Buckinghamshire, as middle class participation in the planning system works to restrict development, thus inflating property prices and excluding low-cost houses and, by extension, reinforcing the middle class character of the community. The result, they note, is,

> a delightful stretch of countryside in the midst of an urban region. With Milton Keynes to the north, green belt and London to the south, the struggle to hold onto the rural in Aylesbury Vale is by no means easy. But the social make-up of the place means that a formidable array of actors can usually be assembled to orchestrate

opposition to unwelcome development. As its positional status grows, the area will become even more attractive to those would-be residents who are trapped on the 'outside'. Thus competition for resources, notably housing, will continue to increase, making it more and more difficult for those on low incomes either to stay in, or move to, such areas. The middle class complexion of the locality is thus assured – at least in the short term – especially as political action and planning policy are likely to reinforce and reflect the prevailing social composition. (Murdoch and Marsden, 1994, p. 229).

The reproduction of middle class space is less complete and less entrenched outside the south east of England, but elements of the same process can be seen in opposition to housing development in areas of the 'contested countryside' (Marsden et al., 1993), such as Somerset. Here, the rhetoric of planning conflicts frequently focuses on the impact of the landscape or local wildlife or on the loss of village character, because such entities are components in the cultural product in which middle class in-migrants have invested. Moreover, there may also be a second tier of, usually unvoiced, motivations for opposition to new development, that include the protection of property prices and concerns about the 'wrong sort' of people moving in who might disrupt the socially aspirational qualities of the rural idyll.

Neither is it only through opposition to significant housing development that middle class investment in the countryside is protected. Campaigns against new roads, industrial developments, quarries and mineral workings, communication masts and windfarms may also follow from a similar rationale. These campaigns are often articulated through a language of environmental protection, contributing to the forging of a coalition in recent year between middle class activism and the broader environmentalist movement. As well as giving legitimacy to the defence of middle class interests, this coalition has also enabled local campaigners to enroll a wider range of skills, tactics and expertise, increasing their capacity to act. An example of this can be seen in the case of opposition to the construction of the M3 extension around Winchester.

Local middle class residents had campaigned against the road since it was first proposed in the early 1970s, motivated by concerns at the landscape impact, noise pollution and the effect on property prices. At first the campaign reflected the middle class composition of the protesters – the professional and technical expertise of the campaigners and their contacts was drawn upon to mount legal challenges and raise technical objections; whilst the old boys network of Winchester School was mobilized to attract political support (Bryant, 1996). These tactics were successful in delaying the road and forcing a revision of the route. The new route, unveiled in 1983, involved taking the road through a cutting in Twyford Down, east of Winchester, and was supported by the local councils. It was opposed, however, by many of the original campaigners and newer middle class in-migrants, who regarded it as despoiling the landscape. Again, the campaign drew on middle class professional skills and contacts, but it also sought to build a network of influential organizations in lobbying for the cutting to be replaced by a tunnel through Twyford Down. A public inquiry was granted in 1987, which in

1990 ruled against the objections. In the final stages of the campaign the continuation of a lobbying and legal strategy was complemented by the adoption of direct action tactics, working with environmental campaigners. In early 1992, local protesters and campaigners from Friends of the Earth occupied water meadows on the route of the road, the blending of the two campaigns symbolized by a the hosting of black tie dinner party on the meadows. The coalition was also joined by radical eco-protesters from the 'Dongas Tribe' of travellers, who occupied part of the proposed construction site and engaged in sabotage of construction machinery until they were forcibly evicted on 'Yellow Wednesday', 9 December 1992, after which the road building commenced (see Bryant, 1996; Lowe and Shaw, 1993; McKay, 1996; also Merrick, 1996 and Parker, 2002 on the Newbury by-pass protests).

Environmentalism versus Preservationism? The Case of Windfarms

Since the start of the twenty-first century the construction of wind turbine power stations has supplanted road-building and house-building as the most visible focal point for campaigning against development in the countryside. In contrast to road-building and house-building, however, windfarms have a claim to be 'environmentally-friendly' developments and have the support of at least part of the green lobby. As such, the issue of windfarm development has undermined the coalition that had been forged in the 1990s between middle class rural campaigners, environmental groups and eco-protesters, and produced new alliances in local rural conflicts. The development of wind power is part of Britain's commitment to increasing its renewable energy production under international agreements. This is managed through the Non-Fossil Fuel Obligation (NFFO), which designates bands of electricity supply for various non-fossil fuel sources whilst seeking to promote competition for contracts to generate this capacity. Thus, contracts for wind energy projects have been awarded under the NFFO in periodic tranches (Horne, 1996). The number of new windfarm projects has increased as concern about the pace of Britain's conversion to renewable energy has grown. Between the opening of the first commercial windfarm, in Cornwall, in 1991 and the beginning of 2004, 86 windfarms had been constructed in Britain. A further 22 were planned to start construction in 2004, with around 20 more reported to be at a preliminary stage in April 2004 (Ward and Dodd, 2004).

The strategy of developing wind energy through competition and private sector investment, combined with geographical limitations on the availability of suitable environmental conditions, means that Britain's wind power capacity has generally been developed through large-scale, commercial, windfarms, most of which have no connection to local communities. This in turn has helped to provoke opposition to windfarm developments, both from local residents and from campaigners outside the localities concerned. Indeed, there is a perception that some early windfarm developments, particularly in Wales, enjoyed the support of local residents but were strongly opposed by external actors. Gipe (1996), for example, notes that the public inquiry for the Cemmaes windfarm in mid Wales in 1992 received statements of support for the development from local councils and the farming unions, whilst opposition was led by the Countryside Council for Wales (a government body), and the Campaign for the Protection of Rural Wales (CPRW) (the Welsh equivalent of the CPRE). However, greater local opposition has been apparent for later windfarm proposals in Wales, Cumbria, Devon, Scotland and elsewhere (Hull, 1995; Woods, 2003b). Increasingly, opponents to windfarms have not only lodged objections through the planning system, but have also adopted direct action and campaigning tactics from other rural protesters. One village in Carmarthenshire, for example, 'changed' its name for a week to a what would have been the longest place name in Britain in order to draw media attention to a proposed wind turbine development nearby. National co-ordination has been orchestrated by the Country Guardian pressure group, initially established in Cumbria in 1991, which claims to have defeated or delayed up to 89 per cent of proposals for windfarms in some years (Barnett and Townsend, 2004).

Protests against windfarms escalated in 2003 and 2004 as the number of new applications also increased. Celebrity endorsements and support from the Conservative party leadership have attracted media attention, but also accusations of links to the nuclear industry and 'middle class NIMBYism' (see for example, Toynbee, 2004). Such depictions, however, oversimplify the complexities of the windfarm debate and in particular the way in which a divide between environmentalism and preservationism has become apparent, with both sides claiming 'green' credentials. The difference lies in how 'nature' is perceived, and the scale at which 'the environment' is imagined. Windfarm supporters tend to represent the environment in global terms, emphasizing problems of climate change, but understand individual local environments as being resilient to changes such as the installation of wind turbines. Opponents of windfarms, in contrast, highlight the vulnerability of local nature.

These polar positions can be seen in the example of the Cefn Croes windfarm, 25 kilometres east of Aberystwyth in the Cambrian Mountains of Wales, on which construction started in 2004. At the time of the planning application in 2000, the Cefn Croes was the largest proposed windfarm in Europe, with 39 turbines of up to 99.7 metres in height (see Woods 2003b for more on the case study). The proposal split the local community, including the local green movement. The development was supported by Friends of the Earth and the former local MP, Cynog Dafis – who had been co-sponsored by the Green party when first elected in 1992 – and opposed by the CPRW, the Countryside Council for Wales, the Open Spaces Society and several local conservation groups. The local Green party narrowly voted to oppose the windfarm following a public meeting to debate the issue. Supporters of the scheme mobilized arguments that represented the environment in global terms, linking the windfarm development to the challenge of climate change and suggesting that the local community had an obligation to back the development as their contribution to tackling the problem:

> Mr Wilson in his letter... asks whether the Cefn Croes Wind Power Station will have a significant effect on global warming. The answer is that it will effect a very significant reduction in Ceredigion's contribution to the problem of global warming

which can only be solved if every community in the world makes a similar reduction. (Letter to the *Cambrian News*, 17 August 2000).

It will produce clean, green electricity without polluting the atmosphere or leaving a dangerous legacy of waste for our children. It is an environment-friendly project that will help Wales to become self-sufficient for its energy needs. I strongly urge people to consider recent footage of the floods in Mozambique, Southern Australia and Venezuela – all probable results of global warming – before considering their position with regard to this project. (Windfarm developer, in letter to the *Cambrian News*, 8 March 2000).

In contrast, the opponents of the Cefn Croes scheme argued that global environmental responsibility meant that individuals had to start by protecting their own local environments and that 'natural areas' such as Cefn Croes therefore needed to protected both for their own intrinsic value and as a part of global biodiversity. The environmental value of Cefn Croes was constructed by campaigners through reference to perceived threats to local wildlife, particularly birds such as the black grouse and red kite; but mostly through reference to the 'natural beauty' of the landscape as well as to attributes such as 'tranquility' which were associated with nature:

If you go up to the top of Pumlumon Fawr and look south, you can see all the way to the Brecon Beacons, you can see all around Cardigan Bay, it's a profoundly beautiful experience to go and stand on there. And to look south into what will just be an industrial mess will be a great loss, it will take a chunk out of my heart I can tell you. (Chair, Cefn Croes Campaign, interview, July 2001).

For those who appreciate the wild and deserted quality of the Cefn Croes area the introduction of a wind farm into its midst will change its character in an unacceptable way. (Ceredigion County Council Planning Department report, quoted in the *Western Mail*, 10 July 2001).

Notably, the arguments of the anti-windfarm campaign have a resonance with the arguments employed by the preservationist movement in the early twentieth century. In both cases the aesthetic quality of the rural landscape is emphasized and perceived urban or industrial intrusions are opposed because of their visual incompatibility with the scale, texture and morphology of the countryside. Brittan (2001) argues that opponents perceive wind turbines as 'mechanical weeds', out of scale with their surroundings, out of harmony with the natural landscape, and potentially out of control. In the Cefn Croes case, campaigners used diagrams to compare the height of the turbines to local landmarks and employed language that described the turbines as an 'industrial mess', 'grit in the eye' and 'a virtual fairground'. Similarly, and ironically, the pro-windfarm lobby reproduces elements of the rural modernization discourse. Windfarms are presented as part of 'progress' towards a more sustainable future, and technological innovation is emphasized. Scientific arguments are mobilized to rationally justify wind power developments, whilst the aesthetic value of landscape is dismissed by analyses of potential sites that are framed by geographical and meteorological considerations.

The rational arguments employed by windfarm developers have more purchase in the planning system than those of the anti-windfarm campaigners, which tend to be more emotive. The pro-windfarm lobby has also been favoured by support from government which, concerned about the impact on its renewable energy strategy of delays to windfarm developments, has indicated a willingness to simplify the planning process for such schemes. Wind power advocates within government are also suggested to be lobbying the Ministry of Defence which had blocked 40 per cent of windfarm proposals between 2000 and 2004 as potential obstacles to low-level training by the Royal Air Force (*Cambrian News*, 2 September 2004). Yet, in spite of this official backing for wind energy, there is a clear sense at the time of writing in late 2004 that public opinion on windfarms could be at a turning point. It is entirely conceivable that within the near future, windfarms will pass the threshold of public acceptability of development in the countryside, joining reservoirs, quarries, power stations and motorways as land uses that are either no longer proposed for rural locations, or which can only be forced through against considerable public opposition.

Conclusion: From Rural Politics to the Politics of the Rural

Contemporary conflicts over the development of the countryside illustrate the thesis that underlies this book, that there has been in Britain a transition from a 'rural politics' to a 'politics of the rural'. Strategies for the development of rural space and proposals for specific developments have always been political and have always attracted debate and conflict. Yet, historically the debates have been framed around technical issues and questions of environmental impact and economic gain that may have had some reference to the rural location but were often not intrinsically about the impact on the perceived rurality of the locale (some preservationist arguments are an obvious exception to this observation). Today, however, questions of rurality are central to conflicts over developments, and concerns about the impact on rurality can over-ride other rational arguments. Similarly, whereas agricultural politics were once essentially industrial politics that happened to have a particular rural spatial expression, contemporary debates and conflicts over the future of farming are intrinsically concerned with questions about the place of agriculture in rural identity. Perhaps most explicitly, the legislative threat to fox-hunting has been presented as a threat to a rural way of life and to the character of the countryside. Groups which have taken up these causes, including most notably the Countryside Alliance and militant protest groups such as the Countryside Action Network, but also arguably Farmers for Action and Country Guardian, can therefore be identified as part of an emerging rural social movement, motivated by a perceived (and not always consistent) rural identity (see Woods, 2003a; 2004).

The tactics adopted by these campaign groups also throw light on the reconfiguration of power relations in rural Britain. Indeed, these can be represented as passing through three phases since the start of the twentieth century. In the early twentieth century, power over the countryside was essentially devolved to a landowning elite who governed rural society through a combination of public and private mechanisms, maintaining their position through a mixture of resource power, associational power and discursive power (Chapter two). Following the First World War, however, there commenced a process of the nationalization of rural politics through which the nation state became the primary power centre. During this period, rural Britain was governed through an external/internal division of labour, with the responsibility of governance within the countryside taken by agricultural and business-led elites operating through the structures of local government, and representation of rural interests to national government being advanced by the twin tracks of established interest groups, such as the NFU, working within closed policy communities, and the Conservative party monopolizing rural constituency representation in parliament.

Over the course of the last quarter-century there has been a denationalization of rural politics in Britain, as in many other countries. At one level this has meant the re-scaling of power downwards to regional and local governance and upwards to the European Union, the World Trade Organization and other supra-national bodies – all of which have created new opportunities for intervention in the political process. At another level, denationalization has meant a decoupling of the rural political process from the institutions of the nation state, with an emphasis on active citizenship, partnership working and the growing power of private sector corporations (see also Woods, 2005c). The targeting of supermarkets and oil companies by rural protesters is a direct reflection of this shift in power. Moreover, as power has been distributed away from the nation state, so the ability of those established rural interest groups who were embedded in state institutions to delivery results to their constituencies has also been undermined. The resulting collapse of confidence in such organizations has been a key factor in the emergence of new, often more radical, groups with a willingness to explore new methods of getting their message across.

The politics of the rural is hence a far more fluid politics than the old-style 'rural politics'. There are a plethora of different groups involved, with differing degrees of organization, pursuing sometimes interlocking but often conflicting goals, and targeting different centres of power. There is a fluidity of movement between scales as local actors build networks and coalitions to engage with actors at higher tiers, and there are wider attempts to construct new networks of power that have a capacity to act across different arenas of rural governance, bringing new actors into the rural political process with new resources to offer. Yet, at the heart of it all remains the concept of rurality – framing policies, informing decision-making, guiding representations and motivating actors. Thus at the start of the twenty-first century, as the contesting of rurality continues to define the terrain of political engagement in rural Britain, it seems that the real power in the British countryside is the very idea of rurality itself.

Bibliography

Aaronovitch, D. (1998), 'Arcadia comes to the Big Smoke, to tell its well-worn tale of woe', *The Independent*, 27 February 1998, p. 21.
Abram, S., Murdoch, J. and Marsden, T. (1996), 'The social construction of 'Middle England': the politics of participation in forward planning', Journal of Rural Studies, Vol. 12, pp. 353-364.
Agyeman, J. and Spooner, R. (1997), 'Ethnicity and the rural environment', in P. Cloke and J. Little (eds), *Contested Countryside Cultures*, Routledge, London and New York, pp. 197-217.
Allison, L. (1975), *Environmental Planning: a political and philosophical analysis*, Allen and Unwin, London.
Anderson, B. (1991), *Imagined Communities*. Verso, London.
Anderson, I. (2002), *Foot and Mouth Disease: Lessons to be Learned – Inquiry Report*, HC888, The Stationery Office, London.
Anon. (1908), *Somersetshire and Some Neighbouring Records: Historical, Biographical and Pictoral*, Allan North, London.
Aslet, C. (1991), *Countryblast: Your countryside needs you*, John Murray, London.
Astor, G. (1961), *The High Sheriff*, Times Publishing Company, London.
Barnes, A.J.L. (1994), 'Ideology and factions', in A. Seldon and S. Ball (eds) *Conservative Century: The Conservative Party since 1900*, Oxford University Press, Oxford, pp. 315-345.
Barnett, A. & Townsend, M. (2004), 'Why Ingham wants to stop wind farms', *The Observer*, 25 April 2004, p. 4.
Barron, J., Crawley, G. and Wood, T. (1991), *Councillors in Crisis*, Macmillan, Basingstoke.
Beard, M. (1989), *English Landed Society in the Twentieth Century*, Routledge, London.
Beckett, A. (1998), 'Yeomen get marching orders', *The Guardian – The Week*, 21 February 1998, p. 1.
Beckett, J. V. (1986), *The Aristocracy in England 1660-1914*, Blackwell, London.
Bickerstaff, K. and Simmons, P. (2004) 'The right tool for the job? Modeling, spatial relationships, and styles of scientific practice in the UK foot and mouth crisis', *Environment and Planning D: Society and Space*, Vol. 22, pp. 393-412.
Birch, A. H. (1959), *Small Town Politics*, Oxford University Press, Oxford.
Body, R. (1982), *Agriculture: The triumph and the shame*, Temple Smith, London.
Bowen Rees, I. (1994), 'Government by Community 1894-1994', in Aitchison, J., Baptiste, J., Jones, P. and Edwards, W. J., *Community and Town Councils of Wales: A Handbook*, Aberystwyth, Rural Surveys Research Unit and Wales Association of Community and Town Councils.
Bracey, H. E. (1959), *English Rural Life*, Routledge and Kegan Paul, London.
Brierley, P. and Hiscock, V. (1993), *UK Christian Handbook 1994/95*, Christian Research Alliance, London.

Brittan, G. G. (2001), 'Wind, energy, landscape: reconciling nature and technology', *Philosophy and Geography*, Vol. 4, pp. 169-184.
Bryant, B. (1996), *Twyford Down: roads, campaigning and environmental law*, Spon, London.
Buller, H. and Lowe, P. (1982), 'Politics and class in rural preservation: a study of the Suffolk Preservation Society', in M. J. Moseley (ed.) *Power, Planning and People in Rural East Anglia*, Centre of East Anglian Studies, Norwich, pp. 21-42.
Bunce, M. (1994) *The Countryside Ideal: Anglo-American images of landscape*, Routledge, London.
Bush, R. (1988), *A Taunton Diary, 1787-1987*, Barracuda Books, Buckingham.
Cahill, K. (2001), *Who Owns Britain*, Canongate, Edinburgh.
Campbell, D. (2004), 'Grey skies over green pleasant land', *The Guardian*, 8 May 2004, p. 9.
Cannadine, D. (1992), *The Decline and Fall of the British Aristocracy*, Picador, London.
Carrington, N. (1939), The Shape of Things to Come, London.
Carter, H. (1998), 'Morale dilemmas', *The Guardian – Society*, 12 August 1998, pp. 2-3.
Charlesworth, A. (ed.) (1983), *An Atlas of Rural Protest in Britain, 1548-1900*, Croom Helm, London.
Clark, G. (1991), 'People working in farming: the changing nature of farmwork', in T. Champion and C. Watkins (eds), *People in the Countryside*, Paul Chapman, London, pp. 67-83.
Cloke, P. (1983), *An Introduction to Rural Settlement Planning*, Methuen, London.
Cloke, P. (1990), 'Community development and political leadership in rural Britain', *Sociologia Ruralis*, Vol. 30, pp. 305-322.
Cloke, P. and Goodwin, M. (1992), 'Conceptualizing countryside change: from post-Fordism to rural structured coherence', *Transactions of the Institute of British Geographers*, Vol. 17, pp. 321-336.
Cloke, P. and Little, J. (1990), *The Rural State?* Oxford University Press, Oxford.
Cloke, P., Milbourne, P. and Widdowfield, R. (2000), 'Partnership and policy networks in rural local governance: homelessness in Taunton', *Public Administration*, Vol. 78, pp. 111-133.
Cohen, A. P. (1985), *The Symbolic Construction of Community*, Routledge, London.
Cook, J. (2001), *The Year of the Pyres: The 2001 Foot and Mouth Epidemic*, Mainstream, Edinburgh.
Countryside Agency (2003), *State of the Countryside 2003*, Countryside Agency, London.
Countryside Alliance (2003), *Annual Report 2002*, Countryside Alliance, London.
County Councils Association (1939), *The Jubilee of County Councils 1889-1939*, Evans Brothers, London.
Cox, K. R. (1998), 'Spaces of dependence, spaces of engagement and the politics of scale, or: looking for local politics', *Political Geography*, Vol. 17, pp. 1-24.
Criddle, B. (1994), 'Members of Parliament', in A. Seldon and S. Ball (eds) *Conservative Century: The Conservative Party since 1900*, Oxford University Press, Oxford, pp. 145-168.
Curtice, J. and Steed, M. (1982), 'Electoral choice and the production of government', *British Journal of Political Science*, Vol. 12, pp. 249-298.
Daniels, S. (1993), *Fields of Vision: Landscape imagery and national identity in England and the United States*, Polity Press, Cambridge.
della Porta, D. and Diani, M. (1999), *Social Movements: an Introduction*, Blackwell, Oxford.
Doherty, B., Paterson, M., Plows, A., Wall, D. (2003), 'Explaining the fuel protests', *British Journal of Politics and International Relations*, Vol. 5, pp. 165-173.

Donaldson, A., Lowe, P. and Ward, N. (2002), 'Virus-crisis-institutional change: the foot and mouth actor network and the governance of rural affairs in the UK', *Sociologia Ruralis*, Vol. 42, pp. 201-214.

Douglas, R. (1976), *Land, People and Politics: A history of the land question in the United Kingdom, 1878-1952*, Allison and Busby, London.

Dunn, P. (1991), 'Villagers bugged by parish-pump 'plot'', *The Independent*, 20 April 1991, p. 3.

Eastwood, D. (1994), *Governing Rural England: Tradition and transformation in local government 1780-1840*, Oxford University Press, Oxford.

Edwards, B., Goodwin, M. and Woods, M. (2003), 'Citizenship, community and participation in small towns: a case study of regeneration partnerships', in R. Imrie and M. Raco (eds), *Urban Renaissance: New Labour, community and urban policy*, Policy Press, Bristol, pp. 181-204.

Edwards, B. and Woods, M. (2004), 'Mobilizing the local: community, participation and governance', in L. Holloway and M. Kneafsey (eds) *Geographies of Rural Cultures and Societies*, Ashgate, Aldershot, pp. 173-198.

Edwards. O. (1989), *A Gem of Welsh Melody*, Ruthin, Coelion Trust.

Elcock, H. (1975), 'English local government reformed: The politics of Humberside', *Public Administration*, Vol. 53, pp. 159-166.

Etzioni-Halevy, E. (1993), *The Elite Connection*, Polity Press, Cambridge.

Evered, P. (1902), *Staghunting with the Devon and Somerset*, Chatto & Windus, London, and James G. Commin, Exeter.

Everett, N. (1994), *The Tory View of Landscape*, Yale University Press, New Haven, CT and London.

Fainstein, S. S. and Hirst, C. (1995), 'Urban social movements', in D. Judge, G. Stoker and H. Wolman (eds), *Theories of Urban Politics*, Sage, London, pp. 181-204.

Flyvbjerg, B. (1998), *Rationality and Power: Democracy in practice*, Chicago, Chicago University Press.

Forbes, I. (2004), 'Making a crisis out of a drama: the political analysis of BSE policy-making in the UK', *Political Studies*, Vol. 52, pp. 342-357.

Fox, G. (1978), *A Country Diary*, G. Fox, Wellington.

Frouws, J., (1998), 'The contested redefinition of the countryside: an analysis of rural discourses in the Netherlands', *Sociologia Ruralis*, Vol. 38, pp. 54-68.

Gaskell, E. (1906), *Somersetshire Leaders: Social and political*, Queenhithe Printing and Publishing, London.

George, J. (1999), *A Rural Uprising? The battle to save hunting with hounds*, Allen and Unwin, London.

Gerlach, L. (1976), 'La Struttura dei Nuovi Movimenti di Rivolta', in A, Melucci (ed), *Movimenti di Rivolti*, Etas, Milan, pp. 218-232.

Gilg, A. (1984), 'Politics and the countryside: the British example', in G. Clark, J. Groenendijk and F. Thissen (eds) *The Changing Countryside*, Geo Books, Norwich.

Gilg, A. (1999), *Perspectives on British Rural Planning Policy, 1994-97*, Ashgate, Aldershot.

Gipe, P. (1995), *Wind Energy Comes of Age*, John Wiley, New York.

Girouard, M. (1978), *Life in the English Country House: A social and architectural history*, Yale University Press, New Haven and London.

Girouard, M. (1981), *The Return to Camelot: Chivalry and the English gentleman*. Yale University Press, New Haven and London.

Grant, W. (1977), *Independent Local Politics in England and Wales*, Saxon House, Farnborough.

Grant, W. (1990), 'Rural politics in Britain', in P. Lowe and M. Bodiguel (eds), *Rural Studies in Britain and France*, Bellhaven, London, pp. 286-298.
Grant, W. (2000), *Pressure Groups and British Politics*, Macmillan, Basingstoke.
Grant, W. (2004), 'Pressure politics: the changing world of pressure groups', *Parliamentary Affairs*, Vol. 57, pp. 408-419.
Green, D. (1996) *Communities in the Countryside*, Social Market Foundation, London.
Griffiths, C. (1999), 'When farmers grumble', *History Today*, Vol. 49, pp. 14-16.
Gruffudd, P. (1995), 'Remaking Wales: nation-building and the geographical imagination, 1925-50', *Political Geography*, Vol. 14, pp. 219-239.
Gummer, J. (1998), 'Those Four Million Houses', in A. Barnett and R. Scruton (eds) *Town and Country*, Jonathan Cape, London.
Guttsman, W. L. (1963), *The British Political Elite*, second edition, MacGibbon and Kee, London.
Hague, W. (1998), 'Marching for freedom', *The Daily Telegraph*, 23 February 1998.
Halfacree, K. (1992), *The Importance of Spatial Representations in Residential Migration to Rural England in the 1980s*, unpublished PhD thesis, Lancaster University.
Halfacree, K. (1996), 'Out of place in the countryside: travellers and the 'rural idyll'', *Antipode*, Vol. 29, pp. 42-71.
Hallam, O. (1971), *The National Farmers Union in Somerset: A history of the county branch 1912-1962*, Somerset NFU, Taunton.
Hanbury-Tenison, R. (1997), 'Life in the countryside', *The Geographical*, November 1997, pp. 88-95 (sponsored feature).
Harper, S. (1988), 'Rural reference groups and images of place', in D. Pocock (ed.) *Humanistic Approaches in Geography*, University of Durham, Department of Geography, Occasional Publication 22.
Hart, S. (2002), 'Liberty, livelihood: a noble cause, if ever there was one', *Liberty and Livelihood March Commemorative Magazine*, pp. 32-39.
Hart Davies, D. (1997), *When The Country Went To Town*, Excellent Press, Ludlow.
Harvey, G. (1998), *The Killing of the Countryside*. Vintage, London.
Heald, T. (1983), *Networks*, Hodder and Stoughton, London.
Heath, A. and Savage, M. (1995), 'Political alignments within the middle class, 1972-89', in T. Butler and M. Savage (eds.) *New Theories of the Middle Class*, UCL Press, London, pp. 275-292.
Hedworth Whitty, R. G. (1934), *The Court of Taunton in the 16th and 17th Centuries*. Goodman and Son, Taunton.
Hetherington, P. (2000), 'Town grouse and country grouse', *The Guardian*, 16 February 2000, p. 17.
Hinchcliffe, S. (2001), 'Indeterminacy indecisions – science, policy and politics in the BSE crisis', *Transactions of the Institute of British Geographers*, Vol. 26, pp. 182-204.
Hinsliff, G. (2000), 'Hunt lobby scents blood at election', *The Observer*, 18 June 2000, p. 11.
Hoggart, K. and Paniagua, A. (2001), 'What rural restructuring?', *Journal of Rural Studies*, Vol. 17, pp. 41-62.
Hooson, E. and Jenkins, G. (1965), *The Heartland: a plan for Mid-Wales*, Liberal Publication Department, London.
Horne, B. (1996), *The Windfarms Debate: the case for and against*, Centre for Alternative Technology, Machynlleth.
Howkins, A. (1991), *Reshaping Rural England: a social history, 1850-1925*, Routledge, London.
Hubbard, P. (2005), 'Inappropriate and incongruous: opposition to asylum centres in the English countryside', *Journal of Rural Studies*, in press.

Bibliography

Hull, A. (1995), 'New models for implementation theory: striking a consensus on windfarms', *Journal of Environmental Planning and Management*, Vol. 38, pp. 285-306.

Ilbery, B. and Bowler, I. (1998), 'From agricultural productivism to post-productivism', in B. Ilbery (ed) *The Geography of Rural Change*, Addison Wesley Longman, Harlow, pp 57-84.

Independent Farmers Group (2002), 'Farmers seek new group to tackle real issues behind farm crisis', press release, available at www.farm.org.uk.

Jackson, J. (2002), 'We would support the lawbreakers', *The Guardian*, 23 October, p. 20.

Jasanoff, S. (1997), 'Civilization and Madness: the great BSE scare of 1996', Public Understanding of Science, Vol. 6, pp. 221-232.

Johnson, R. W. (1972), 'The nationalisation of English rural politics: Norfolk South West 1945-70', *Parliamentary Affairs*, Vol. 26, pp. 8-55.

Johnston, R. J., Pattie, C. J. and Allsopp, J. G. (1988), *A Nation Diving?* Longman, London.

Jowell, R., Witherspoon, S., Brook, L. and Taylor, B. (eds) (1990), *British Social Attitudes: the seventh report*, Gower, Aldershot.

Karpf, A. (1998), 'March FM: You could hear the duke's millions in the jingles alone', *The Guardian*, 2 March, p. 5.

Latour, B. (1986), 'The powers of association', in J. Law (ed) *Power, Action and Belief: A new sociology of knowledge*, Routledge, London, pp. 264-280.

Leach, J. (1998), 'Madness, metaphors and miscommunication: the rhetorical life of Mad Cow Disease', in S. C. Ratzen (ed.) *The Mad Cow Crisis: Health and public good*, UCL Press, London, pp. 119-130.

Leaney, S. (2001) *Diary of a Farming Wife: One women's view of the foot and mouth crisis*, Merton Priory Press, Cardiff.

Lee, J. M. (1963), *Social Leaders and Public Persons*, Oxford University Press, Oxford.

Lowe, P. and Goyder, J. (1983), *Environmental Groups in Politics*, Allen and Unwin, London.

Lowe, P., Murdoch, J. and Cox, G., (1995), 'A civilised retreat? Anti-urbanism, rurality and the making of Anglo-centric culture', in P. Healey, S. Cameron, S. Daroudi, S. Graham and A. Madani-Pour (eds) *Managing Cities: the new urban context*, Wiley, London, pp. 63-82.

Lowe, P., Murdoch, J., and Norton, A. (2001), *Professionals and Volunteers in the Environmental Process*, Centre for Rural Economy, Newcastle.

Lowe, P., Murdoch, J., Marsden, T., Munton, R. and Flynn, A. (1993), 'Regulating the new rural spaces: the uneven development of land', *Journal of Rural Studies*, Vol. 9, pp. 205-222.

Lowe, R. and Shaw, W. (1993), *Travellers: Voices of the New Age Nomads*, Fourth Estate, London.

Lusoli, W. and Ward, S. (2003), 'Hunting protestors: mobilisation, participation and protest online in the Countryside Alliance', paper presented to the ECPR Joint Sessions, University of Edinburgh, April.

Luttrell, C. (1925), *The Sporting Recollections of a Younger Son*, Duckworth, London.

Macnaghten, P. and Urry, J. (1998), *Contested Natures*, Sage, London and Thousand Oaks.

Madgwick, P.J., Griffiths, N., and Walker, V. (1974), *The Politics of Rural Wales*, Hutchinson, London.

Marsden, T., Murdoch, J., Lowe P., Munton, R., and A. Flynn (1993), *Constructing the Countryside: An approach to rural development*, UCL Press, London.

Marsh, D. and Rhodes, R. (eds) (1992), *Policy Networks in British Governance*, Oxford University Press, Oxford.

Matless, D. (1990), 'Definitions of England, 1928-89: Preservation, modernism and the nature of the nation', *Built Environment*, Vol. 16, pp. 179-191.

Matless, D. (1994), 'Doing the English Village, 1945-90: An essay in imaginative geography', in P. Cloke, M. Doel, D. Matless, M. Phillips and N. Thrift, *Writing The Rural: Five cultural geographies*, Paul Chapman, London, pp. 7-88.

Matless, D. (1998), *Landscape and Englishness*, Reaktion Books, London.

May, T. and Nugent, N. (1982), 'Insiders, outsiders and thresholders: corporatism and pressure group strategies in Britain', paper presented at the Political Studies Association conference, University of Kent.

McKay, G., (1996), *Senseless Acts of Beauty*, Verso, London and New York.

Meacher, M. (2001), 'Gummer is wrong', Letter to *The Guardian*, 9 July, p. 15.

Merrick (1996), *Battle for the Trees*, Godhaven Ink, Leeds.

Milbourne, P., (2003), 'The complexities of hunting in rural England and Wales', *Sociologia Ruralis*, Vol. 43, pp. 289-308.

Miller, D. (1999), 'Risk, science and policy: definitional struggles, information management, the media and BSE', *Social Science and Medicine*, Vol. 49, pp. 1239-1255.

Mingay, G. (ed.) (1989), *The Unquiet Countryside*, Routledge, London.

Mitchell, G. D. (1951), 'The Parish and the Rural Community', *Public Administration*, Vol. 29, p. 397.

Mitchell, J. V. and Dolun, M. (2001), *The Fuel Tax Protests in Europe, 2000-2001*, Royal Institute of International Affairs, London.

Monbiot, G. (1998), 'Conservation by rights', in A. Barnett and R Scruton (eds) *Town and County*, Jonathan Cape, London, pp. 91-98.

Mormont, M. (1987), 'The emergence of rural struggles and their ideological effects', *International Journal of Urban and Regional Research*, Vol. 7, pp. 559-78.

Mormont, M. (1990), '"What is rural?" or 'How to be rural': towards a sociology of the rural', in T. Marsden, P. Lowe, and S. Whatmore (eds), *Rural Restructuring*, David Fulton, London, pp. 21-44.

Murdoch, J. (1992), 'Representing the region: Welsh farmers and the British state', in T. Marsden, P. Lowe and S. Whatmore (eds), *Labour and Locality: Uneven development in the labour process*, David Fulton, London, pp. 160-181.

Murdoch, J. (1997), 'The shifting territory of government: some insights from the Rural White Paper', *Area*, Vol. 29, pp. 109-118.

Murdoch, J. and Abram, S. (2002), *Rationalities of Planning: development versus environment in planning for housing*, Ashgate, Aldershot.

Murdoch, J. and Lowe, P. (2003), 'The preservationist paradox: modernism, environmentalism and the politics of spatial division', *Transactions of the Institute of British Geographers*, Vol. 28, pp. 318-332.

Murdoch, J. and Marsden, T. (1994), *Reconstituting Rurality*, UCL Press, London.

Murdoch, J. and Marsden, T. (1995), 'The spatialization of politics: local and national actor-spaces in environmental conflict', *Transactions of the Institute of British Geographers*, Vol. 20, pp. 368-380.

Moore, C. (1998), 'We're on the march', *The Daily Telegraph*, 2 March, p. 20.

Moore, S. (1991), 'The agrarian Conservative Party in parliament, 1920-1929', *Parliamentary History*, Vol. 10, pp. 342-362.

Mosbacher, M. and Anderson, D. (eds) (1999), *Another Country*, Social Affairs Unit, London.

Moylan P.A. (1978), *The Form and Reform of County Government: Kent 1889-1914*. Occasional Papers, 3rd series, no 3, Department of English Local History, University of Leicester, Leicester.

Newby, H. (1977), *The Deferential Worker*, Allen Lane, London.
Newby, H. (1987), *Country Life: A social history of rural England*, Cardinal, London.
Newby, H., Bell, C., Rose, D., and Saunders, P. (1978), *Property, Paternalism and Power*, Hutchinson, London.
North, R. A. E. (2001), *The Death of British Agriculture*, Duckworth, London.
Offe, C. (1985), 'New Social Movements: changing boundaries of the political', *Social Research*, Vol. 52, pp. 817-868.
Packett, C. N. (1975), *The County Lieutenancy in the United Kingdom (1547-1975)*, published by the author, Bradford.
Park, A. et al. (2001), *British Social Attitudes: the eighteenth report*, Sage, London.
Parker, G. (2002), *Citizenships, Contingency and the Countryside: Rights, culture, land and the environment*, Routledge, London.
Pattie, C., Dorling, D. and Johnston, R. (1995), 'A debt-owning democracy: the political impact of housing market recession on the British General Election of 1992', *Urban Studies*, Vol. 32, pp. 1293-1315.
Paxman, J. (1991), *Friends in High Places*, Penguin, London.
Paxman, J., (1998), *The English: A portrait of a people*, Michael Joseph, London.
Phillips, P. W. B. (1990), *Wheat, Europe and the GATT*, Pinter, London.
Phythian-Adams, C. (1972), 'Ceremony and the citizen: the communal year at Coventry 1450-1550', in P. Clark and P. Slack (eds) *Crisis and Order in English Towns, 1500-1700*, Routledge and Kegan Paul, London.
Phythian-Adams, C. (1989), 'Rural culture', in G.E. Mingay (ed.), *The Vanishing Countryman*, Routledge, London, pp. 76-86.
Porter, H. (1997), 'Another country', *The Guardian – Section 2*, 10 July, pp. 2-3.
Press, C. A. M. (1890), *Liberal Leaders of Somerset*, Simpkin, Marshall, Hamilton, Kent & Co., London and John Whitby & Son, Bridgwater.
Press, C. A. M. (1894), *Somersetshire Lives: Social and political*, Gaskell, Jones & Co, London.
Private Eye (2001), *Not the Foot and Mouth Report*, Pressdram, London.
Ramet, S., (1996), 'Nationalism and the 'idiocy' of the countryside: the case of Serbia', *Ethnic and Racial Studies*, Vol. 19, pp. 70-86.
Ratzen, S. C. (ed.) (1998), *The Mad Cow Crisis: Health and public good*, UCL Press, London.
Redlich, J. and Hirst, F.W. (1958), *The History of Local Government in England*, Macmillan, London.
Reed, M. (2004), 'The mobilisation of rural identities and the failure of the rural protest movement in the UK, 1996-2001', *Space and Polity*, Vol. 8, pp. 25-42.
Rentoul, J. (2001), *Tony Blair: Prime Minister*, Little, Brown and Company, London.
Richards, H. (1996), 'The view from within', *Farming Wales*, October, p. 15.
Robinson, N. (2003), 'Fuel protests: governing the ungovernable?', *Parliamentary Affairs*, Vol. 56, pp. 423-440.
Rose, D., Saunders, P., Newby, H., and Bell, C. (1976), 'Ideologies of property: a case study', *Sociological Review*, Vol. 24, pp. 699-730.
Routledge, P. (1997), 'The imagineering of resistance: Pollock Free State and the practice of postmodern politics', *Transactions of the Institute of British Geographers*, Vol. 22, pp. 359-376.
RPM (2002), *The Rich at Play: Foxhunting, land ownership and the 'Countryside Alliance'*, Revolutions Per Minute, London.
Rural Wales (2002), 'Foot and Mouth Disease 2001: The impact on small businesses in Powys', *Rural Wales*, Summer, pp. 14-15.

Saunders, P., Newby, H., Bell, C. and Rose, D. (1978), 'Rural community and rural community power', in H. Newby (ed.), *International Perspectives in Rural Sociology*, John Wiley and Sons, London, pp. 55-86.

Saville, J. (1957), *Rural Depopulation in England and Wales, 1851-1951*, Routledge and Kegan Paul, London.

Schindegger, F. and Krajasits, C. (1997), 'Commuting: its importance for rural employment analysis', in R. D. Bollman and J. M. Bryden (eds), *Rural Employment: an International Perspective*, CAB International, Wallingford, pp. 164-176.

Scott, A., Christie, M. and Midmore, P. (2004), 'Impact of the 2001 foot-and-mouth outbreak in Britain: implications for rural studies', *Journal of Rural Studies*, Vol. 20, pp. 1-14.

Serow, W., (1991), 'Recent trends and future prospects for urban-rural migration in Europe', *Sociologia Ruralis*, Vol. 31, pp. 269-280.

Sharp, T. (1946), *The Anatomy of the Village*, Penguin Books, Harmondsworth.

Sheingate, A. D. (2001), *The Rise of the Agricultural Welfare State*, Princeton University Press, Princeton, NJ.

Sheppard, H. B. (1909), *Courts Leet and the Court Leet of the Borough of Taunton*, Barnicott and Pearce, Taunton.

Short, J. R. (1991), *Imagined Country: Environment, culture and society*, Routledge, London.

Sibley, D., (1997), 'Endangering the sacred: nomads, youth cultures and the English countryside', in P. Cloke and J. Little, (eds), *Contested Countryside Cultures*, Routledge, London and New York, pp. 218-231.

Sinclair, G. (1992), *The Lost Land: Land use change in England, 1945-1990*, CPRE, London.

Smith, M. J. (1989), 'Changing agendas and policy communities: Agricultural issues in the 1930s and the 1980s', *Public Administration*, Vol. 67, pp. 149-65.

Smith, M. J. (1993), *Pressure, Power and Policy*, Harvester Wheatsheaf, Hemel Hempstead.

Stacey, M. (1960), *Tradition and Change: A study of Banbury*, Oxford University Press, London.

Stanyer, J. (1975), 'Farming politics: Devon and Cheshire compared', in D. Murray (ed.) *Decision Making in Britain III: Agriculture, parts 6-9*. Open University: Milton Keynes.

Stanyer, J. (1989), *The History of Devon County Council, 1888-1988*, Devon Books, Exeter.

Stone, C. (1988), 'Pre-emptive power: Floyd Hunter's 'Community Power Structure' reconsidered', *American Journal of Political Science*, Vol. 32, pp. 82-104.

Sutton, P. (2000), *Explaining Environmentalism: in search of a new social movement*, Ashgate, Aldershot.

Taylor, D. M. (1996), 'Bovine spongiform encephalopathy – the beginning of the end?', *British Veterinary Journal*, Vol. 152, pp. 501-518.

Thomas, G. (1996), 'Crisis in the countryside', *The Western Mail*, 22 March.

Thompson, F. M. L. (1963) *Landed Society*, Routledge and Kegan Paul, London.

Thrift, N. (1989), 'Images of social change', in C. Hamnett, L. McDowell and P. Sarre (eds.) *The Changing Social Structure*, Sage, London, pp. 12-42.

Thrift, N. (1996), *Spatial Formations*. Sage, London.

Toynbee, P. (2004),'Countryside alliance', *The Guardian*, 13 August, p. 23.

Tregidga, G. (2000), *The Liberal Party in South West Britain since 1918*, University of Exeter Press, Exeter.

Trollope, J. (1998), 'The country we love', *The Sunday Telegraph - Review*, 1 March, p. 1.

Turner, W. J. (1947), *Exmoor Village*, George Harrap, London.

Urry, J. (1995), 'A middle-class countryside?', in T. Butler and M. Savage (eds.) *New Theories of the Middle Class*, UCL Press, London, pp. 205-219.

Bibliography

Vogel, U. (1989), 'The land question: A liberal theory of communal property', *History Workshop*, Vol. 27, pp. 106-135.
Vidal, J. (1998), 'Masters of illusion', *The Guardian – Society*, 4 March, pp. 4-5.
Walker, J. L. (1991), *Mobilizing Interest Groups in America*, University of Michigan Press, Ann Arbor.
Walton, E. (2002), 'They can't lock us all up, can they?', *The Field*, August, pp. 48-52.
Ward, D. and Dodd, V. (2004) 'Storm of protest over planned windfarm', *The Guardian*, 5 July, p. 24.
Ward, N. (2002), 'Representing rurality? New Labour and the electoral geography of rural Britain', *Area*, Vol. 34, pp. 171-181.
Ward, N., Donaldson, A. and Lowe, P. (2004), 'Policy framing and learning the lessons from the UK's foot and mouth disease crisis', *Environment and Planning C: Government and Policy*, Vol. 22, pp. 291-306.
Webb, S. and Webb, B. (1963), *The Parish and the County: English Local Government, Volume I*, Frank Cass & Co. Ltd, London.
Welsh Office (1964), *Depopulation in Mid Wales*, HMSO, London.
Whatmore, S. (2002), *Hybrid Geographies: Natures, cultures, spaces*, Sage, London and Thousand Oaks.
White, R. J. (ed.) (1950), *The Conservative Tradition*, N. Kaye, London.
Will, R. G., Ironside, J. W., Zeidler, M., Cousens, S. N., Estibeiro, K., Alperovitch, A., Poser, S., Pocchiari, M., Hofman, A., Smith, P. G. (1996), 'A new variant of Creutzfeldt-Jakob disease in the UK', *The Lancet*, No. 347, pp. 921-925.
Williams, R. (1973), *The Country and the City*, Chatto and Windus, London.
Williams-Ellis, C. (1928) *England and the Octopus*, Geoffrey Bles, London.
Winter, M. (1996a), *Rural Politics*, Routledge: London and New York.
Winter, M. (1996b), 'Intersecting departmental responsibilities, administrative confusion and the role of science in government: the case of BSE', *Parliamentary Affairs*, Vol. 49, pp. 550-565.
Woods, M. (1997), 'Discourses of power and rurality: local politics in Somerset in the 20th century', *Political Geography*, Vol. 16, pp. 453-478.
Woods, M. (1998a), 'Advocating rurality? The repositioning of rural local government,', *Journal of Rural Studies*, Vol. 14, pp. 13-26.
Woods, M. (1998b), 'Rethinking elites: networks, space and local politics', *Environment and Planning A*, Vol. 30, pp. 2101-2119.
Woods, M. (1998c), 'Researching rural conflicts: hunting, local politics and actor-networks', *Journal of Rural Studies*, Vol. 14, pp. 321-340.
Woods, M. (1998d), 'Mad cows and hounded deer: political representations of animals in the British countryside', *Environment and Planning A*, Vol. 30, pp. 1219-1234.
Woods, M. (1999), 'Performing power: local politics and the Taunton pageant of 1928', *Journal of Historical Geography*, Vol. 25, pp. 57-74.
Woods, M. (2002), 'Was there a rural rebellion? Labour and the countryside vote in the 2001 General Election', in L. Bennie, C. Rallings, J. Tonge and P. Webb (eds) *British Elections and Parties Review: Volume 12 – The 2001 General Election*, Frank Cass, London, pp. 206-228.
Woods, M. (2003a), 'Deconstructing rural protest: the emergence of a new social movement', *Journal of Rural Studies*, Vol. 19, pp. 309-325.
Woods, M. (2003b), 'Conflicting environmental visions of the rural: windfarm development in Mid Wales', *Sociologia Ruralis*, Vol. 43, pp. 271-288.
Woods, M. (2004), 'Politics and protest in the contemporary countryside', in L. Holloway and M. Kneafsey (eds), *The Geographies of Rural Societies and Cultures*, Ashgate, Aldershot, pp. 103-125.

Woods, M. (2005a), *Rural Geography*, Sage, London and Thousand Oaks.
Woods, M. (2005b), 'Redefining rural affairs as a policy area', in J. Loughlin and M. Bogdani (eds), *Policy Convergence and Divergence in a Devolving United Kingdom: Rural Affairs Policy*, Policy Press, Bristol, forthcoming.
Woods, M. (2005c), 'Political articulation: the modalities of the new rural citizenship', in P. Cloke, T. Marsden and P. Mooney (eds) *The Handbook of Rural Studies*, Sage, London and Thousand Oaks.
Young, K. (1994), 'The party and English local government', in A. Seldon and S. Ball (eds.) *Conservative Century: The Conservative Party since 1900*, Oxford University Press, Oxford, pp. 403-443.
Young, K. (1988), 'Interim report: Rural prospects', in Jowell, R., Witherspoon, S., and Brook, L. (eds) *British Social Attitudes: the fifth report*, Gower, Aldershot, pp. 155-174.
Young, K. and Mills, L. (1990), 'The governance of rural England', Appendix G in The Archbishops' Commission on Rural Areas, *Faith in the Countryside*, ACORA Publishing, Stoneleigh Park.

Index

Abercrombie, Patrick 171, 172
Abram, Simone 178, 180
access to countryside 3, 21
Acland-Hood family 27, 30, 72
agrarian countryside, discourse of 3, 131, 140, 171
agricultural elite 12, 33-5, 36, 42-5, 46, 47, 84, 85, 131, 167, 168, 170, 186; *see also* elites; elites networks
agricultural workers 24, 30, 34, 44, 45, 90, 110, 119, 121, 123, 132
agriculture 7, 12, 15, 34, 37, 42, 51, 83, 86, 89, 98, 100, 107, 111, 123, 124, 131–61, 162, 164, 165, 167, 191
  crisis in 2, 3, 99, 104, 121, 141–5, 146, 152
  employment 15, 17, 43, 44, 99
  policy 12, 17, 86, 93, 94, 95, 111, 131–7, 144, 158-60, 163
  subsidies 17, 93, 121, 132, 136, 142–3
  Ministers of 85, 98, 145
Agriculture Act (1948) 41, 132
aldermen 10, 26, 29, 36, 46, 73, 74, 84
Anglesey, Isle of 146–8
apolitical countryside, the 3–6, 25–6, 37, 45, 119, 120
aristocracy, the 8, 10, 12, 23, 30, 31–2, 50, 72, 84, 89, 93, 122, 170
Aylesbury Vale 178–81, 186

Baldwin, Stanley 89, 169–70
Berkshire 119, 184, 185
Bettelheim, Eric 101, 124
Birmingham 107, 129
Blair, Tony 97, 98, 99, 125, 155, 157
Bracey, H. E. 33, 41, 42
Bridgwater 26, 30, 51
British Field Sports Society 101–3, 122, 124
British National Party 90
Bruford, Robert 36, 85
BSE 18, 137, 139, 140, 153
  and CJD 18, 137, 139, 141
  crisis 93, 137–40, 141, 142, 143, 159
  export ban on beef 138, 139, 140, 142, 146
  as a rural crisis 141-2
Buckinghamshire 20, 50, 122, 178–81, 186-7
Burge, Richard 122, 125
business elite 11, 33, 35–6, 41, 46, 51, 64, 75–7, 83, 84, 168; *see also* elites; elite networks
Butler, Sam 102, 103, 122, 125

Cambridgeshire 178, 184, 185
Campaign for the Protection of Rural Wales 189
Campaign to Protect Rural England, *see* CPRE
CAP, *see* Common Agricultural Policy
capitalism 12, 36, 45, 48, 165, 166
Cardiff 107, 125, 144, 145, 148, 154
Cardiganshire 2, 23, 96, 167, 168; *see also* Ceredigion
Carmarthenshire 99, 145, 146, 189
Cefn Croes windfarm 189–91
Cely-Trevilian, Maurice 32, 38–9, 40
Ceredigion 189; *see also* Cardiganshire
chambers of trade/commerce 11, 36, 54, 175, 177
Cheshire 31–2, 34, 138, 150, 156, 184
civic societies 75, 77, 181
Clark family 29, 70, 73, 80
class 8, 21, 29-30, 37, 39, 40, 44, 47–8, 85, 86, 91, 92, 94, 95, 96, 108, 110, 122, 167, 186
clergy 8, 26, 35
Cloke, Paul 47, 48, 50, 165
Comité des Organizations Professionelles Agricoles (COPA) 134, 136, 159
Common Agricultural Policy 93, 134, 135–6, 150; *see also* agriculture – policy
  reforms 136, 159

community 7—9, 35, 37, 39, 40, 87, 179, 186
  leadership 35, 37, 41, 42, 43, 46, 49, 61
  organic community 8, 36—7, 87, 88
Confederation Paysanne 108, 149
Conservatism 36-7, 49, 84—94, 101, 119, 169—70
  and rurality 5, 78, 85-94, 146
Conservative party 36, 37, 46, 48, 51-2, 54, 57, 65, 67, 70, 75, 77-8, 84, 87, 88, 90, 91, 92, 94—103, 105, 114, 122, 125, 129, 134, 170, 179, 184, 189
  and farmers 34, 43, 85-6, 93, 95, 131, 146
Cornwall 2, 95, 104, 117, 138, 188
Council for the Preservation/Protection of Rural England, *see* CPRE
councillors 23, 26-7, 29, 32, 33, 35, 41, 43, 46, 49—50, 52, 53, 54—7, 61, 62, 64, 67, 70, 72, 73, 75, 80, 84, 88, 91, 167, 176, 181, 183
counterurbanization 2, 3, 15, 17, 18, 42, 45, 47-8, 51, 94, 95, 98, 168, 173, 176, 177, 186
Country Business Group 101—2, 103, 122, 124
country gentleman, discourse of the 3, 27—9, 34, 37, 38, 83, 170, 171
Country Guardian 189, 191
Country Landowners Association (CLA) 12, 67, 84, 85—6, 131, 132, 133, 181
Countryside Action Network 3, 18, 127, 130, 191
Countryside Agency 67, 156
Countryside Alliance 2, 3, 18, 98, 103—30, 162, 175, 191; *see also* Countryside March, Countryside Rally, Liberty and Livelihood March
  activities 107, 110
  finance 105, 122
  internal conflict 124—5
  leadership 122—3
  membership 110, 121—2
  organization 105, 107
  origins 103, 124
Countryside Council for Wales 189
Countryside March (1998) 2, 18, 104, 112, 114, 115, 117, 119, 121, 122, 123, 124, 125, 185; *see also* Countryside Alliance, Countryside Rally, Liberty and Livelihood March
Countryside Movement 102, 103, 122, 124
Countryside Protection Group 121
Countryside Rally (1997) 2, 18, 103—4, 109, 112, 114, 115, 117, 119, 123, 124; *see also* Countryside Alliance, Countryside March, Liberty and Livelihood March
county councils 7, 8, 11, 19, 23, 26-7, 30, 32-3, 33—4, 36, 37, 39, 40, 41, 43, 45, 92, 167, 171
CPRE 11, 84, 101, 121, 170, 171—6, 177, 178, 181, 182, 183, 184, 185, 189
Criminal Justice and Public Order Act (1994) 87—8
cultural capital 7, 30, 48, 49, 53
Cumbria 2, 99, 104, 122, 156, 157, 189

DEFRA 109, 157-8; *see also* Department of the Environment; MAFF
Department for the Environment, Food and Rural Affairs, *see* DEFRA; DETR
Department of the Environment 176, 177; *see also* DEFRA
Department of the Environment, Transport and the Regions, *see* DETR
DETR 179, 184, 184
deputy lieutenants 10, 27, 32, 67—70, 72
devolution 144, 154, 159
Devon 27, 33, 95, 138, 152, 156, 176, 184, 189
discourses of power, *see* power, discourses of
discourses of rurality, *see* rurality, discourses of
district councils 43, 45, 64
Duke, Edward 125, 128
Dumfries 99, 156
Dunster 29, 30, 32, 72
Durham 2, 99, 156

East Anglia 42-5, 47, 167
Edinburgh 107, 122, 127, 138, 154
EEC, *see* European Community
elections 23, 59, 69, 73, 88, 94—7, 144, 152, 184

elite networks 25, 29-31, 32, 34, 36, 44, 49, 54–76; see also agricultural elite; business elite; elites
elites 5–11, 17, 25, 27, 30, 33-6, 39, 40, 46, 48, 49, 50, 54, 64, 65, 79, 82, 84, 86, 92, 109, 164, 167, 168, 176; see also agricultural elite; business elite; elite networks
England 85, 88, 89-90, 93, 94, 95, 120, 121, 123, 132, 133, 139, 144, 151, 156, 159, 164, 166, 167, 169, 177, 178, 181, 185–7
environmentalism 14, 83, 136, 140, 149, 163, 170, 174, 187, 188–91
Essex 152, 185
European Commission 134, 135, 136, 138, 139, 152; see also European Community
European Community 17, 93, 134, 135, 136, 138, 139, 140, 159, 192; see also European Commission
European Union, see European Community
Exeter 98, 99, 107
Exmoor 51, 63, 72, 122
  national park 51, 72, 73

'Farm' (farming union) 160
farmers 6, 36, 86, 95, 96, 104, 110, 122, 129, 132, 134–7, 139–42, 146–52, 157, 159–60, 162–3, 174
  in local politics 26, 33–5, 37, 40, 41–5, 46, 49, 50, 52–3, 131
  in parliament 85, 93
  protests 21, 99, 126, 145, 146–52, 159, 160, 161
  as representatives of countryside 131
  response to BSE 139–40, 141–2
  tenant farmers 8, 24, 85
Farmers for Action 3, 18, 148–9, 150, 151, 159–61, 191
Farmers' Union of Wales (FUW) 105, 126, 134, 140, 141, 144, 145, 146, 149, 159, 161
Farming, see agriculture
farming unions 12, 93, 131–4, 136, 140, 143–4, 145, 146, 148, 158, 159, 189; see also Farmers' Union of Wales; National Farmers' Union.
feudalism 7, 11, 72, 83

First World War 12, 24, 31, 34, 43, 45, 85, 131, 132, 169, 192
Foot and Mouth disease 152, 153
  1967 outbreak 152
  closure of footpaths 153, 155
  cost 153, 155, 156
  crisis 3, 18, 21, 99, 100, 104, 127, 152-8, 159
  eradication strategy 152–4
  export ban 152–3
  impact on farmers 153, 154, 155–6
  Lessons to be Learned Inquiry 155
France 2, 108, 139, 146, 149, 159
Freemasons 11, 36, 54, 64, 75-7
Friends of the Earth 78, 183, 185, 188, 189
fuel protests (2000) 3, 18, 21, 99, 126, 149–52

Gass, Lady Elizabeth 63, 67, 72
gender 28, 39, 121
gentry 5, 8, 10, 12, 23, 30, 31–2, 33, 35, 41, 43, 50, 84, 89, 93, 170
George, Janet 102, 104, 124-5, 127
Germany 137, 150
Gladstone, William 12, 30
globalization 15, 111, 162
Gloucestershire 156, 157, 184
Goldsmith, Zac 160
Grant, Wyn 35, 131, 136, 159, 161
Green party 182, 183, 189
greenbelt 164, 168, 177, 178, 184, 186
Gwyther, Christine 145

Hague, William 93, 114, 116–17
Hampshire 150, 184
Hanbury-Tennison, Robin 102
Hanley, David 148, 149, 150, 151, 152
health authorities 60, 64, 77
Herefordshire 85, 152, 156
Hertfordshire 176, 184
High Sheriff 7, 10, 29, 32, 38, 39, 50, 67–9, 70, 72, 74
Hobhouse family 26–7, 30, 32
Hogg, Quintin 86–7
Holyhead 146, 148
Howkins, Alun 31, 33
Housebuilders' Federation 168, 177, 181
housing development 48, 123, 164, 176–86

politics of 3, 12, 49, 94, 99, 104, 111, 123, 177–86
hunting 9, 101, 121
   and local power 29-30, 36, 54, 62, 64
   supporters 101–3, 104, 111, 113, 120, 123, 124–6, 144
   threat of a ban 2, 21, 61, 83, 94, 98, 99, 101–3, 104, 107, 111, 113–15, 121, 123, 124, 125–6, 127, 129, 152, 191
Hunting Declaration 125

In-migrants 1, 2, 16, 18, 42, 45, 46, 47–50, 51, 57, 64, 83, 95, 96, 111, 122, 174, 176, 177, 182, 186, 187
Incomers, *see* in-migrants

Jackson, John 122, 125, 126
Jarrow March 117, 119, 121
Justices of the Peace, *see* magistrates

Kimball, Lord 122

Labour party 5, 6, 37, 41, 42, 46, 48, 52, 60, 61, 65, 78, 80, 82, 83, 85, 94, 96–100, 102, 107, 111, 114, 115, 122, 123, 124, 128, 145, 146, 148, 154, 172, 183, 185
   New Labour 97, 98–100, 154
land reform 24, 31
landed elite 9, 10, 11, 12, 24, 31, 33, 47, 83, 84, 93, 101–2, 122, 123, 129, 170
landed gentry, *see* gentry
landowners 6, 7, 8, 10, 12, 26, 27, 32, 34, 36, 37, 38, 39, 43, 44, 45, 47, 62, 70, 84, 85, 96, 104, 110, 129, 132, 170, 174
landownership 7, 8, 10–11, 29, 44, 86, 111
Lee, J M 31-2, 34
Liberal Democrats 46, 48, 51, 52, 57, 65, 70, 78, 80, 83, 88, 94, 95-6, 97, 102, 179, 180, 184; *see also* Liberal party; Liberal SDP Alliance
Liberal party 24, 26, 31, 36, 60, 85, 94, 95, 96, 102, 134, 167; *see also* Liberal Democrats, Liberal SDP Alliance

Liberal SDP Alliance 46, 52, 94, 95; *see also* Liberal party, Liberal Democrats
Liberty and Livelihood March 2–3, 18, 104–7, 112, 115, 117, 119, 125, 129; *see also* Countryside Alliance; Countryside March; Countryside Rally
Lincolnshire 2, 24, 34, 150
local government 7, 9, 19, 23–4, 26–7, 33, 34, 41–3, 45-6, 48, 50, 52-3, 54–63, 64, 65, 67, 72, 73–4, 77, 79, 81–2, 83, 84, 92, 94, 96, 131, 166–7, 168, 170, 172, 173, 176, 185, 186
localism 38–9, 40, 49, 87
London 6, 51, 65, 101, 103, 104, 105, 107, 109, 112, 115, 121, 122, 129, 146, 148, 154, 161, 177, 178, 184, 186
Lord Lieutenant 7, 9, 23, 32, 39, 50, 54, 67–70, 72
lord of the manor 7, 67, 74; *see also* squire
Lowe, Philip 164, 172, 174
Luttrell family 29, 30, 32, 63, 67, 72, 80

mad cow disease, *see* BSE
Madgwick, Peter 2, 96, 168
MAFF 12, 86, 132, 133, 134, 136, 137, 152, 153–4, 155, 156, 157
magistrates 7, 27, 29, 30, 32, 38, 43, 50, 57, 61, 64, 67, 69, 72, 73, 75, 82, 84, 174
Major, John 93, 94, 138
Mallalieu, Baroness Anne 114, 117, 122, 123
March on the Mound 107, 122
market towns 11, 35–6, 48, 97, 167, 171, 177
Marsden, Terry 20, 50, 186
Members of Parliament 8, 12, 26–7, 29, 30, 32, 36, 39, 42, 67, 72, 74, 84, 85, 86, 87, 93, 94, 97, 107, 113, 125, 128
Michael, Alun 145
middle class 2, 42, 45, 47–8, 50, 51, 62, 64, 73, 83, 94, 95, 165, 176, 186–7, 188
militant pro-hunting groups 3, 18, 125–30
milk quotas 93, 136, 146

Milton Keynes 178, 179, 185, 186
Minehead 51, 54, 63, 64, 82
Ministry of Agriculture, Fisheries and Food, *see* MAFF
Ministry of Defence 120, 191
Moore, Charles 116–17, 119, 124
Monmouthshire 148, 156
Mormont, Marc 18–19, 163
Murdoch, Jonathan 20, 50, 174, 178, 180, 186

National Assembly for Wales 144–5, 152, 154
National Farmers' Union 12, 17, 18, 34–5, 37, 54, 84, 85, 86, 101, 105, 131–4, 136, 139, 140, 143, 144, 146, 148, 149, 153–4, 159–60, 161, 173, 192
national identity 3, 40, 86, 88–91, 115–17, 120–21, 124
  British 90, 116–17, 119, 120
  Cornish 117, 120–21
  English 86, 88, 89–90, 91, 96, 116, 117, 120–21
  Scottish 96, 117, 120–21, 169, 171
  Welsh 96, 117, 120–21, 144
National Trust 20, 48, 121, 169, 175
National Union of Agricultural Workers 6, 133
natural order 8, 29, 171
nature 164, 165, 189, 190
new age travellers 87–8, 188
Newby, Howard 8, 28, 42–5, 47
NFU, *see* National Farmers' Union
NHS Trusts 54, 60, 67, 72, 75
Norfolk 6, 29, 42, 43
North Yorkshire 138, 156; *see also* Yorkshire
Northern Ireland 139, 154, 159
Northumberland 152, 156, 176

organic society 86–8, 89, 91

pageants 3, 38–40
parish 7–9, 33
parish councils 1, 7, 11, 19, 23, 33, 49, 50, 59, 82, 84
paternalism 2, 12, 17, 20, 44, 46, 72, 74, 123, 167
patronage 7, 12, 60–62, 69, 83
Paxman, Jeremy 50, 90

Plaid Cymru 96, 144
planning policy 163–4, 165–6, 168, 177–86, 187
planning system 165, 166, 172, 173, 175, 177, 180, 181
policy communities 17, 18, 20, 21, 132, 192
  agricultural policy community 17, 93, 101, 132–5, 136, 154; *see also* agriculture – policy
political occupations 35, 36, 46, 53
Portman family 30, 31
postmodern politics of resistance 109, 127, 149
power 6–10, 12, 20, 24–6, 27, 29, 30, 32, 33, 34, 35–6, 37, 41–45, 46, 53, 54, 63, 65, 72, 79, 80, 81, 82, 84, 113, 149, 192
  discourses of 25-6, 27-8, 37, 40, 85, 131
Powys 122, 156
preservationist movement 164, 169–76, 177, 188–91
pressure groups 19, 103–5, 125, 132–7, 144
  insider groups 17, 129, 132–3, 159, 161, 173
  thresholder groups 129, 173
productivism 132, 135, 136, 137, 140, 163; *see also* agriculture – policy
property 44, 45, 70, 114; *see also* landownership
  and class 48; *see also* class
  and Conservatism 86–88; *see also* Conservatism
  prices 95, 176–7, 186–7
  rights 36, 85, 86, 87–8, 91, 123

quarter sessions 7, 8, 23
Quantock Hills 51, 63

racism 87, 90, 123
Real Countryside Alliance 3, 18, 127–8, 130
recruitment of political activists 57–60, 83
right to roam 94, 98, 99, 123; *see also* access to the countryside
road building 94, 111, 164, 165, 167, 168, 187–8
Rotary clubs 11, 36, 38, 54, 75, 77, 79

Rural Action Group 125
Rural Coalition (USA) 108, 160
Rural Development Commission 12, 166
rural district councils (RDCs) 9, 33, 36, 43, 45, 73, 167
Rural Group of Labour MPs 98, 99; see also Labour party
rural idyll 3, 18, 45, 47, 48, 49, 89, 111, 116, 166, 179, 182, 186, 187
rural movement 108–9, 111, 129, 191; see also social movements
Rural Rebels 18, 127
rural restructuring 12–20, 21, 89, 95, 116
rural services, closure of 2–3, 98, 104, 124
rural – urban divide 11, 18, 29, 37, 89, 94, 98, 113–15, 122, 170, 171
Rural White Papers 17, 94, 98
rurality, discourses of 3, 11, 26, 27-8, 37, 86, 88, 123, 124, 131, 162, 164, 166, 170, 171

Sanders, Sir Robert (Lord Bayford) 30, 32, 85
school governors 43, 61, 73, 75
Scotland 85, 94, 95, 97, 104, 117, 121, 127, 139, 154, 159, 164, 189
Scottish National Party 96
Scottish Parliament 107, 127, 154
Second World War 41, 46, 132, 135, 158, 172, 179
service class 2, 48–50, 51, 57, 95, 96
Severn Bridge 125, 148
sheriff, see High Sheriff
Social Affairs Unit 94, 168
social movements 108–11, 129, 150
socialism 5, 6, 39, 87
Somerset 1, 26–31, 32, 34, 35, 39, 40, 41, 49, 50, 51–76, 86, 90, 91, 92, 95, 138, 142, 178, 181–4, 185, 187
Somerset County Council 26–7, 29, 30, 32, 33, 34, 40, 41, 51–2, 53, 54, 61, 63, 65, 69, 70, 72, 73, 83, 88, 142, 181, 184
Soroptomists 54, 75, 77
Sportsmans' March 102–3
squire 11, 37, 43; see also lord of the manor; squirearchy
squirearchy 24, 26, 31, 89
Staffordshire 23, 156

Stanlow oil refinery 150, 151
Stanyer, Jeffrey 7, 27
stewardship 27–9, 44, 83, 140, 141–2
Suffolk 42–3, 92, 174, 185
supermarkets 142, 146, 148, 149, 159, 160

Taunton 26, 35, 36, 38–40, 51, 52, 54, 57, 64, 72, 74–77, 79, 83
  Court Leet 54, 74–7, 78
  Pageant 38–40
Taunton Deane 51, 59, 65, 88, 178, 183–4
Taunton Deane Borough Council 51–2, 53, 57, 59, 63, 64, 65, 73, 75, 83, 184
teachers in rural local government 35, 52–3, 54, 64
Thatcher, Margaret 92, 93
Town and Country Planning Act (1947) 132, 172
trades unions 6, 34, 37, 42, 85, 121, 132, 146
training and enterprise councils 54, 75
Trollope, Joanna 116, 117, 123
Twyford Down 187–8

Under Sheriffs 59
Union of Country Sports Workers 125
United Kingdom Independence Party 90
United States of America 2, 12, 108, 137, 146, 159
Urry, John 48, 138

village amenity societies 178, 180

Waldegrave family 30, 72
Wales 23, 85, 94, 95, 96, 104, 117, 121, 132, 133, 134, 142, 143, 144–5, 146–8, 149, 150, 154, 156, 157, 158, 159, 165, 166, 167, 188, 189–91
Warwickshire 151, 161
Watchet 51, 90
Wellington 51, 54, 64, 73
Wells 30, 36, 85
Welsh Assembly, see National Assembly for Wales
West Somerset 51, 72, 79
West Somerset District Council 52, 53, 64, 65, 73
West Sussex 50, 184

Williams, Brynle 149, 150, 151, 152
Williams-Ellis, Clough 169–70
Wills family 29, 67
Wiltshire 138, 152
windfarms 187, 188–91
Winter, Michael 12, 132–3, 173

Worcestershire 23, 156

Yeovil 26, 51
Yorkshire 120, 125, 128; *see also* North Yorkshire